MEDICAL GENETICS: DEVELOPMENT OF ETHICAL DIMENSIONS IN CLINICAL PRACTICE AND RESEARCH

The transcript of a Witness Seminar held by the History of Modern Biomedicine Research Group, Queen Mary University of London, on 14 October 2014

Edited by E M Jones and E M Tansey

Volume 57   2016

First published by Queen Mary University of London, 2016

The History of Modern Biomedicine Research Group is funded by the Wellcome Trust, which is a registered charity, no. 210183.

ISBN 978 1 91019 5130

All volumes are freely available online at www.histmodbiomed.org

Please cite as: Jones E M, Tansey E M. (eds) (2016) *Medical Genetics: Development of Ethical Dimensions in Clinical Practice and Research*. Wellcome Witnesses to Contemporary Medicine, vol. 57. London: Queen Mary University of London.

CONTENTS

WHAT IS A WITNESS SEMINAR?

The Witness Seminar is a specialized form of oral history, where several individuals associated with a particular set of circumstances or events are invited to meet together to discuss, debate, and agree or disagree about their memories. The meeting is recorded, transcribed, and edited for publication.

This format was first devised and used by the Wellcome Trust's History of Twentieth Century Medicine Group in 1993 to address issues associated with the discovery of monoclonal antibodies. We developed this approach after holding a conventional seminar, given by a medical historian, on the discovery of interferon. Many members of the invited audience were scientists or others involved in that work, and the detailed and revealing discussion session afterwards alerted us to the importance of recording 'communal' eyewitness testimonies. We learned that the Institute for Contemporary British History held meetings to examine modern political, diplomatic, and economic history, which they called Witness Seminars, and this seemed a suitable title for us to use also.

The unexpected success of our first Witness Seminar, as assessed by the willingness of the participants to attend, speak frankly, agree and disagree, and also by many requests for its transcript, encouraged us to develop the Witness Seminar model into a full programme, and since then more than 60 meetings have been held and published on a wide array of biomedical topics.[1] These seminars have proved an ideal way to bring together clinicians, scientists, and others interested in contemporary medical history to share their memories. We are not seeking a consensus, but are providing the opportunity to hear an array of voices, many little known, of individuals who were 'there at the time' and thus able to question, ratify, or disagree with others' accounts – a form of open peer-review. The material records of the meeting also create archival sources for present and future use.

The History of Twentieth Century Medicine Group became a part of the Wellcome Trust's Centre for the History of Medicine at UCL in October 2000 and remained so until September 2010. It has been part of the School of History, Queen Mary University of London, since October 2010, as the History of Modern Biomedicine Research Group, which the Wellcome Trust

---

[1] See pages 113–19 for a full list of Witness Seminars held, details of the published volumes, and other related publications.

funds principally under a Strategic Award entitled 'The Makers of Modern Biomedicine'. The Witness Seminar format continues to be a major part of that programme, although now the subjects are largely focused on areas of strategic importance to the Wellcome Trust, including the neurosciences, clinical genetics, and medical technology.[2]

Once an appropriate topic has been agreed, usually after discussion with a specialist adviser, suitable participants are identified and invited. As the organization of the seminar progresses and the participants' list is compiled, a flexible outline plan for the meeting is devised, with assistance from the meeting's designated chairman/moderator. Each participant is sent an attendance list and a copy of this programme before the meeting. Seminars last for about four hours; occasionally full-day meetings have been held. After each meeting the raw transcript is sent to every participant, each of whom is asked to check his or her own contribution and to provide brief biographical details for an appendix. The editors incorporate participants' minor corrections and turn the transcript into readable text, with footnotes, appendices, a glossary, and a bibliography. Extensive research and liaison with the participants is conducted to produce the final script, which is then sent to every contributor for approval and to assign copyright to the Wellcome Trust. Copies of the original, and edited, transcripts and additional correspondence generated by the editorial process are all deposited with the records of each meeting in the Wellcome Library, London (archival reference GC/253) and are available for study.

For all our volumes, we hope that, even if the precise details of the more technical sections are not clear to the non-specialist, the sense and significance of the events will be understandable to all readers. Our aim is that the volumes inform those with a general interest in the history of modern medicine and medical science; provide historians with new insights, fresh material for study, and further themes for research; and emphasize to the participants that their own working lives are of proper and necessary concern to historians.

---

[2] See our Group's website at www.histmodbiomed.org.

ACKNOWLEDGEMENTS

We are very grateful to Professor Peter Harper for suggesting the topic for this seminar, and also for advising us on suitable participants and the intended programme for discussion. Many thanks also to Professor Anneke Lucassen for chairing the meeting so successfully.

As with all our meetings, we depend a great deal on Wellcome Trust staff to ensure their smooth running: the Audiovisual Department, Catering, Reception, Security, and Wellcome Images. We are also grateful to Mr Akio Morishima for the design and production of this volume; the indexer Ms Cath Topliff; Mrs Sarah Beanland and Ms Fiona Plowman for proofreading; Mrs Debra Gee for transcribing the Seminar; Ms Caroline Overy for assisting with running the seminar and Mr Adam Wilkinson who assisted in the organization and running of the meeting. Finally, we thank the Wellcome Trust for supporting the Witness Seminar programme.

Tilli Tansey

Emma Jones

School of History, Queen Mary University of London

ILLUSTRATIONS AND CREDITS*

* Unless otherwise stated, all photographs were taken by Thomas Farnetti, Wellcome Trust, and reproduced courtesy of the Wellcome Library, London.

INTRODUCTION

It is my pleasure to introduce the latest Witness Seminar volume from the History of Modern Biomedicine Research Group at Queen Mary University of London. The topic is one close to my heart, as my first edited book was on the topic of ethics in clinical genetics: *Case Analysis in Clinical Ethics* was a collaboration with the chair of the present seminar, Professor Anneke Lucassen, and with one of the participants, Professor Michael Parker, as well as a distinguished Dutch colleague, Professor Guy Widdershoven.[1]

In the late 1990s and early 2000s clinical genetics was one of the central topics of the then fast developing and professionalizing field of academic bioethics. A number of different factors made clinical ethics exemplary of what contemporary bioethics was trying to do. One key element was the role of research funders. In the UK, for example, the Wellcome Trust decided to broaden its own funding portfolio in 'medicine, society and history' to include biomedical ethics, with a quite explicit agenda to develop not only 'public understanding' of the medical sciences but also primary research into ethical issues. Such research was to be interdisciplinary in method, with a strong focus on policy relevance for society at large but also for the Trust intramurally. The research funded by the Trust was not only to provide specific answers to specific policy questions but also to be a mechanism for capacity building for UK bioethics itself. Many people now active in bioethics in the UK were either trained through funding from the Wellcome Trust for PhDs and research fellowships or clinician training fellowships. Others who entered bioethics later in their careers (including myself) obtained their first research project grants from the Trust. And still others, who may have been active in teaching medical ethics, began to reshape their research and publication activities in order to take advantage of Wellcome funding.

Much of this activity was driven by the ordinary internal processes of discipline formation, but equally there were push factors. As is discussed in the seminar, the General Medical Council began to require medical ethics to be formally taught as part of undergraduate medical education in the early 1990s. But the drive, under the pressure of the Research Assessment Exercises and increasingly stringent university funding conditions, for medical schools to become highly research intensive and to underwrite their high operating costs with increased external research funding led to significant pressure on those employed to teach medical ethics in medical schools to do externally funded research. As none of the Research

---

[1]   Ashcroft *et al.* (eds) (2005).

Councils took medical ethics or bioethics to be 'core business' of their funding streams, the position of the Wellcome Trust was unique, and its influence over the type, form, and topic of research funding of bioethics research was critical.

Although, initially, the Trust was very open about the kinds of topics it was prepared to fund, provided these were practically relevant and assisted in capacity building, at a relatively early stage it decided to prioritize three areas: international research ethics (which was a special area of interest for the Trust given its overseas research centres), neurosciences and ethics, and genetics.[2] Most of the work eventually funded by the Trust in the first ten years or so of its biomedical ethics funding was in the area of genetics, specifically genetic *research*. So, one way to characterize biomedical ethics and its interest in genetics in the late 1990s and early 2000s is to see it as interdisciplinary research into the ethical issues arising from research in genetics, in order to produce evidence for policy. Understood this way, bioethics begins to become quite different from medical ethics as it had been practised in UK universities and medical schools from the late 1960s onwards. Up to that point, medical ethics could be described as fundamentally a pedagogical and pastoral discipline, concerned with 'moral dilemmas', and very much focused on the practical decision-making of clinicians and medical students. However, we also need to notice a third strand in the reshaping of bioethics, equally important in some ways to the other two, but less frequently acknowledged. One reason the Wellcome Trust wanted to emphasize both interdisciplinarity and research was a perception that hitherto most public discussion of bioethical topics – for example, the Warnock Committee on Human Fertilisation and the Embryo (1984) – focused primarily on discussion within small groups of the 'great and the good', and armchair reflection on concepts and intuitions (what is pejoratively referred to as 'Oxford philosophy').[3]

On the one hand these methods were not congenial to the Trust, inasmuch as its main efforts were devoted to research in the sciences and to the role of empirical evidence; then, as now, research funders struggle to understand exactly what it is that philosophers do, and why, and how to tell whether it is any good. On the other hand, there was a persistent unease that the sort of ethical arguments

---

[2]   The Review of the Wellcome Trust Policy Unit (*c.*2001) states 'The Biomedical Ethics Programme was set up as a fixed-term, five-year initiative at the Wellcome Trust in 1997, with a budget of £5 million. The broad aim of the Programme is to support empirical research into the ethical, legal, social, and public policy aspects of developments in biomedical science, with a focus on two areas: neuroscience (including mental health) and genetics.' It also notes that the scope of the programme should be widened to include international research; see pages 4 and 12.

[3]   Committee of Inquiry into Human Fertilisation and Embryology and Warnock (1984).

produced by elite groups might systematically fail to grasp what was of concern to the public at large, or to important sections thereof. This is just a concrete instance of the trend away from the 'public understanding of science' model developed by the Royal Society at the beginning of the 1990s to the broader and more dialogical conception of 'public engagement with science'. It began to be felt quite widely within the scientific establishment that for the governance of science to be legitimate (that is, not only normatively acceptable but actually accepted) the public had to be engaged in genuine two-way dialogue. It is, as they say, no accident that in the earliest days of the biomedical ethics funding stream it was led by a leading science journalist (Tom Wilkie) and a sociologist-activist (Patricia Spallone) who saw biomedical ethics as a way of expanding this kind of engagement. Although this might seem an unusual stance for a major research charity to take, and although the Wellcome Trust's approach to public engagement and biomedical ethics were to change quite rapidly and significantly through the 2000s, and after, this position is explicable if we notice that much of the discussion of ethics and genetics outside the higher education sector was led by activist groups such as Genewatch UK, which though small were very effectively networked in the media. The stances of these groups tend to be shaped by the positions of the 1970s radical science movement, and were in general highly sceptical about the benefits to humankind, which might accrue from genetic science and technologies.[4] Although nothing in human genetics provoked the kind of large-scale and polarized reactions that genetically modified foods did throughout the 2000s, many policy makers and scientists worried that such a reaction might arise. It is this concern, if not with democracy as such, then at least with 'societal risk management', which motivated a lot of activity in bioethics research funding.

Reading over the transcript of the seminar, several of the themes I have described above do emerge, but it is interesting to me how different the geneticists' and clinicians' construal of what was going on was to mine, as it developed over the past 20 or so years. For one thing, unsurprisingly, the clinicians are powerfully engaged by questions of what it is to practise medicine well, how to care for patients when there is little or nothing that can be done to change their condition (as was the case throughout much of the period under review), and how it is this question of 'good medicine' that is central, rather than, say, ethical principle or sociological evidence about what patients and their families wanted, or indeed wider societal issues. Inasmuch as wider societal issues did arise, these were framed very much in terms of moving the conversation away from genetics-as-eugenics to genetics-as-family-medicine or, especially, genetics-

---

[4]   For an overview of the UK's radical science movement, see Bell (2015).

as-counselling. The conception of counselling itself mutates rapidly away from advice-giving towards something more psychotherapeutic in the form of 'non-directive counselling' and helping patients to make their own best decisions. In turn this is seen as being in tension with both what good medicine requires and with what patients actually want from professionals. The question of eugenics seems largely to disappear as not arising in, or salient to, what actually happens in clinic. Instead, two other kinds of difficulty do become central.

One has to do with research. A clinician has care of her patients, and the difficulty of working out how to help patients who no longer come one-by-one, but instead in family groups, with all the troubles that attend all families (except, perhaps, Tolstoy's mythical happy ones – all of which are alike in that they have the common property of non-existence). In addition, for most clinical geneticists, much of the time, the disorders they are called upon to diagnose may be rare, or even previously unknown, such that every patient is actually or potentially a research subject, not only because they are 'interesting material' but in order to do anything for them at all, clinically. The need for cooperation between clinicians, and indeed, often, between families, and also internationally, creates both a research network and a clinical network, with quite different styles of governance and legal and regulatory rules being brought into contact. Consequently, confusion and delay often arises, and out of this mess emerges a call for principles, or advice, where the bioethics community is purportedly expert, even where it is divided and regularly required to make things up as it goes along. (A very good example of this is the tangled web of ethical discussion relating to human tissue before and after the Alder Hey scandal in the early 2000s – compare and contrast the Nuffield Council on Bioethics' reports on human tissue in 1995 and 2011 and the Human Tissue Act in England of 2004).[5]

As we move into the era of the mainstreaming of genomics, the previous regime of clinical genetics focusing on careful and painstaking work to solve puzzles in small numbers of patients known intimately to their clinicians seems wildly different to the world of big data analytics, statistical inference, electronic patient records, and whole genome scanning. This in turn is in a complex and as yet unresolved relationship with the model of genetic governance arising out of genetic epidemiology (for example, the work of ALSPAC/'Children of the Nineties').[6] We

---

[5] Nuffield Council on Bioethics (1995, 2011), and the Human Tissue Act 2004 (*c.*30).

[6] My first project as Principal Investigator was a Wellcome Trust-funded study on 'ethical perspectives on epidemiological genetics: participants' perspectives', involving focus groups and interviews with children and parents involved in the Avon Longitudinal Study of Pregnancy and Childhood (ALSPAC) (1997–2000). For ALSPAC, see also Overy, Reynolds, and Tansey (eds) (2012).

see here both the classic sociological transition from a social organization based on trust and status to one based on rule and contract; and now from one based on rule and contract to one based on… what? Algorithm and code? Time will tell.

The other kind of difficulty is emotional. Several of the participants at the seminar referred to the emotional labour, indeed the emotional burden, of their work. To some extent we can understand the emergent role of the genetic counsellor as an institutional solution to this problem, but although this might transform the location, and some of the modalities, of that emotional labour, it does not displace it entirely. Yet almost no work was done within bioethics on the affective dimension of clinical work and its role in thinking ethically about such work. This is consistent with what happened in medical education in the period, with a transition from a training in professionalism that followed an apprenticeship model – long after this had been abandoned in the medical sciences – to a formalized model requiring training in 'communication skills' alongside, and separated from 'medical ethics and law'. Yet even communication skills training, in figuring the clinical relationship as something reducible to a learnable set of skills, framed by an externally imposed set of ethical and legal regulations, ignored or possibly tried to overcome, the affective relationship central to medical *care*.

In conclusion: in addition to the invaluable historical testimony shared in this seminar, we have an invaluable resource for thinking about what bioethics is for, what good medicine is, and various ways in which regulation and principle displace and transform but do not replace or overcome professionalism, care, emotion, and indeed the embodied selves of both patients and clinicians. Clinical genetics in this sense represents the exemplary medical specialty: doing ethics while doing medicine, in the face of emotional frailty and scientific doubt. Just not, I think, in the way the Wellcome Trust – and bioethicists – thought it was back in the 1990s.

Professor Richard Ashcroft
Queen Mary University of London

Figure A

MEDICAL GENETICS: DEVELOPMENT OF ETHICAL DIMENSIONS IN CLINICAL PRACTICE AND RESEARCH

The transcript of a Witness Seminar held by the History of Modern Biomedicine Research Group, Queen Mary University of London, on 14 October 2014

Edited by E M Jones and E M Tansey

MEDICAL GENETICS: DEVELOPMENT OF ETHICAL DIMENSIONS IN CLINICAL PRACTICE AND RESEARCH

Participants*

Dr Mark Bale

Professor Angus Clarke

Dr Nick Dennis

Professor Bobbie Farsides

Dr Alan Fryer

Dr Nina Hallowell

Professor Peter Harper

Professor Shirley Hodgson

Professor Anneke Lucassen (Chair)

Professor Bernadette Modell

Mrs Elizabeth Mumford

Professor Michael Parker

Professor Marcus Pembrey

Professor Martin Richards

Professor Tilli Tansey

Professor Peter Turnpenny

Apologies include: Dr Caroline Berry, Professor Martin Bobrow, Professor Ruth Chadwick, Professor John Harris, Professor Theresa Marteau, Professor Susan Michie, Professor Heather Skirton, Professor Rosamund Scott, Professor Stephen Wilkinson

* Biographical notes on the participants are located at the end of the volume

Professor Tilli Tansey: Let me introduce myself: I'm Tilli Tansey and I'm Professor of the History of Modern Medical Sciences at Queen Mary University of London, and the convenor of these Witness Seminars, which are opportunities to get behind the written record of modern history, of modern science and medicine, to find out what really happened. Who were the drivers; what did happen; why did it happen? Things were never inevitable; there were usually reasons why things happened or why they didn't happen, so those are the stories we hope that you will tell us about this afternoon, how they developed in relation to the development of ethics and a clear code of practice in research and clinical genetics.

We hope today to get behind the published records, the official records, and the published literature, and, with the advice of Peter Harper, we've devised an outline programme of themes to try to address.[1] This is, however, not a rigid programme. If at any point there are things that are not included that you wish to say, do please say them. This is your opportunity to tell your history.

The founding medical geneticists and ethical issues
Early genetic counselling: The emergence of 'non-directiveness' as a core principle
Recognition of special ethical problems in clinical genetics practice and research: what are they and how unique to genetics?
• Reproductive issues
• Presymptomatic genetic testing
• Confidentiality and consent issues; genetic testing of children
• Population screening
Involvement of social scientists and the humanities in genetics
Development of genetic counsellors and their influence on ethical aspects of practice
Formal bodies involved in ethics and genetics: Nuffield Council on Bioethics, Human Genetics Commission, etc.

Table 1: Witness Seminar outline programme[2]

---

[1]   Professor Peter Harper is a consultant to the History of Modern Biomedicine Research Group on clinical genetics for the 'Makers of Modern Biomedicine: Testimonies and Legacy' project, funded by a Wellcome Trust Strategic Award.

[2]   The outline programme was circulated to participants in advance of the Witness Seminar.

Figure 1: Professor Tilli Tansey, Professor Anneke Lucassen

An important part of any of these meetings is, of course, finding a suitable chairman, and that really is a misnomer because this is not a meeting in which we want some sort of consensus to emerge but really we want to hear the diversity of views, opinions, and experiences. I think it's probably better to use the word facilitator, and we're delighted that Anneke Lucassen has agreed to adopt that role for us. Basically, she's going to keep you all in order and get us to tea and drinks afterwards on time, so thank you very much, Anneke, and over to you.

Professor Anneke Lucassen: Thank you very much. I think I know most of you. I'm an academic clinician in Southampton and I've had a longstanding interest in exploring the ethical issues raised in clinical genetics. And one of the ways I've developed that interest in practice is, over the last 10 years, through a group with Mike Parker, Angus Clarke, who are here today, and Tara Clancy, called the Genethics Club.[3] Apparently 'club' is a bit exclusive, so we call it the Genethics Forum now – this is a meeting for anyone interested to come and talk about practical issues that raise ethical aspects in genetics.

We started off that meeting as a one-off in 2001, thinking that we'd sort out all the ethical issues in one or two meetings, but actually it grew from there and

---

[3]   The Genethics Club (later Forum) was founded in 2001; see Lucassen and Parker (2006), and also on the Ethox Centre's website; www.ndph.ox.ac.uk/research/ethox-centre/ethics-support/genethics-club (accessed 28 July 2015). See also note 152. Dr Tara Clancy is a Consultant Genetic Counsellor and Lecturer in Medical Genetics at the Manchester Centre for Genomic Medicine, Central Manchester University Hospitals NHS Foundation Trust; http://mangen.co.uk/about-us/OurStaff/Consultants/TaraClancy.php (accessed 3 July 2015).

Figure 2: Professor Peter Harper

there's been a continued demand to keep them going, and we've got our 40th meeting next year. I hope to draw on some of that experience.

Today, I'm not going to be chairing, I'm going to be facilitating as you discuss things as they come up, so what I wanted to do was ask Peter Harper to start the ball rolling and then hope that you'll all contribute. I might ask some of you directly to contribute bits and pieces, but please also volunteer if you've got something to say and hopefully with a nice, free-floating structure like that we'll get some good discussion going. Right, I should just say that I'm a newbie to these seminars, and a lot of you, I know, have been before so you'll have to direct me if I'm not doing it quite right. Peter, over to you.

Professor Peter Harper: Anneke and Tilli asked would I start things going by exploring the question: how did ethics and ethical issues, how did they first get into what we might loosely call medical genetics? I think it's useful to do this because some, indeed most, of the founding people in the field, both in this country and in other countries are no longer living and probably quite a few of you won't have ever known them. I've been lucky in the sense that I'm old enough to have known quite a few of them and so are some other people here. I think one has to start from the perspective that up until the end of World War II the whole area of human genetics was in very bad standing following the abuses of eugenics, which had really gone on for the first half of the twentieth century.[4] This had, in

---

[4] See, for example, Harper (2008), chapter 15, 'Eugenics', pages 405–27.

fact, alienated a lot of people, a lot of scientists, from taking part in the field. Yet my feeling, which people can argue with, is that since that time, over the past 50 years, ethics has become very strongly embedded in the ethos and the practice of medical genetics, perhaps particularly clinical genetics. The question is: how did it get like that? It might perhaps have gone the other way.

So, thinking of where to start, actually a good person to start with, though I did not know him well, and there are people here who can say much more, is Lionel Penrose at London's Galton Laboratory. After he came back from Canada, I think it was in 1945, and took the Galton Chair, which was originally called the Chair of Eugenics, but he managed to get that and his department's name changed, he really set the standard for something quite new. [5] He, I think it's fair to say, confronted the previous situation pretty well head-on, and you can see that in the inaugural lecture that he gave, in which he used phenylketonuria as an example.[6] But I think it was a lot wider than that, and one of the things that I've been very impressed with while interviewing early people in the field around the world, doing recorded interviews, is, first of all, actually, the number of people who got their initial training with him.[7] Not just people in this country but people from continental European countries – for instance, people like Jean Frézal in France, Jan Mohr in Copenhagen, and a number from North America like Arno Motulsky and Barton Childs.[8] I've been very impressed, not just by

---

[5] Professor Lionel Sharples Penrose (1898–1972) was Galton Professor of Human Genetics from 1945 to 1965 at University College London (UCL); for further biographical details see Harris (1973). For the Galton Laboratory's history at UCL, see Jones (1993), and also Harper (2008), pages 235–40.

[6] Penrose (1946).

[7] Transcripts of Professor Peter Harper's interviews with leading medical geneticists for the Genetics and Medicine Historical Network, based at Cardiff University, are freely available to download at https://genmedhist.eshg.org/39.0.html (accessed 11 January 2016).

[8] Dr Jean Frézal (1922–2007) was a geneticist based at the Hôpital Necker, Paris, where he developed medical genetics research and clinical services. Jan Mohr (1921–2009) was the Head of the Institute of Medical Genetics at the University of Oslo, then Professor of Human Genetics at Copenhagen University from 1964 to 1991; see http://icmm.ku.dk/klinikken/the_clinic/history/jan_mohr/ (accessed 7 July 2015). See Professor Peter Harper's interviews with Frézal and Mohr for the Genetics and Medicine Historical Network at https://genmedhist.eshg.org/39.0.html (accessed 11 January 2016). Dr Arno Motulsky developed medical genetics at the University of Washington, where he is now Professor Emeritus of Medicine and Genome Sciences; see further biographical details at http://depts.washington.edu/medgen/faculty/Arno_Motulsky.shtml (accessed 7 July 2015). Professor Barton Childs MD (d. 2010) was the first Director of Genetics in the Department of Pediatrics at Johns Hopkins University, Baltimore; see an obituary at www.hopkinschildrens.org/barton-childs-obituary.aspx. (accessed 7 July 2015).

the number of key people who were very much influenced by training with him but by the tremendous respect in which they held him. I think, actually, this goes directly to this word 'respect' because it's fair to say Penrose had a great deal of respect for the subjects he was working with who, for the most part, were people who were mentally handicapped. That's a striking contrast to what had happened before the war, in many cases, where really there was no respect for subjects, and these individuals were often classed as subhuman or just viewed as problems. So there was that and, of course, he had great respect for the facts too in terms of the rigorous science.

All the people I've interviewed who have worked with him, unhesitatingly, have named him when I've asked who has been the greatest influence. Just about all of them have given his name if they have worked with him, so I think he had a huge influence, but, on the other hand, I don't feel that Lionel Penrose was himself a clinical geneticist. He was clinical, very much so, and he was a geneticist, but clinical genetics hadn't really taken shape at that point and it wouldn't until the next generation. So his role was to set the ethos, as you might say, and to mould the people who would themselves be the first practising clinical geneticists. When we think about it now, who were those people?

I think clinical medical genetics really first got off the ground in an organized way not in Britain but in North America, and there are three key people I'd like just to mention who I think paved the way for it there: these are Clarke Fraser in Montreal, who started a unit there about 1950; Arno Motulsky in Seattle; and, of course, Victor McKusick in Baltimore, both of whom founded departments or divisions of departments in 1957.[9] They, in turn, trained large numbers of people, and I think it's this fact more than their actual research that stands out and has made them iconic figures throughout the world; not just North America because perhaps especially Victor McKusick was responsible for training people in clinical genetics from, really, everywhere, including a considerable number from this country.[10] Then the people who had been trained by McKusick, Fraser, and Motulsky went on to train many other people themselves and formed really a considerable body of people in the early years of medical genetics.

---

[9] Dr (Frank) Clarke Fraser (1920–2014) founded the Human Genetics Unit at the Montreal Children's Hospital in 1952, affiliated to McGill University's Genetics Department; see http://biology.mcgill.ca/fraser. html. For Victor McKusick, see biography on page 88–9.

[10] Professor Peter Harper also trained with Victor McKusick.

I've been trying to think, were these people particularly conscious of ethical issues and was there any particular aspect of practice that stands out? To be honest, I don't think they were particularly aware, certainly they weren't aware in any formal sense of ethics as a philosophical field.[11] They were all good clinicians, experienced clinicians, and they were all what I would consider to be principled people, but I don't think they were especially conscious of ethics as a particular field. I think that also goes for the people in Britain who were the founders: Cyril Clarke in Liverpool, Paul Polani at Guy's, Cedric Carter and John Fraser Roberts at Great Ormond Street.[12] Again, most of them, well three out of those four, were experienced clinicians. Fraser Roberts, I seem to remember Marcus Pembrey saying, he qualified in medicine and then hung up his stethoscope, and that was a reasonable thing to do because he wouldn't have got the respect perhaps and the entrée without his medical qualification. But I don't think any of them were formally conscious of ethical aspects. Cedric Carter, of course, was a proponent of eugenics but I actually don't think that crept into his practice particularly. I think that he was, he behaved as, a principled person and I think this was why all these people were very well respected.

I suppose really, just to leave things there, I don't think ethics came in consciously at that early stage except, and it's a very important 'except', that I think the core of people who founded the field of medical genetics were principled people, kindly people also, who were respected both by the people they saw in terms of patients and families, and by the people who trained and worked with them. This gave the area a sound foundation and a foundation, which was very much needed after what had happened before, and it's that which I hope will be picked up on. Of course, you could say there weren't very many practical issues arising in those early years that raised ethical aspects. That all lay in the future, and it will be interesting to wonder how they might have dealt with these. But it was a sound foundation, and I think one that needs to be recognized and I hope other people will be able to elaborate on that.

Lucassen: Thank you, Peter. Can I just clarify, you say you don't think those founding fathers were really involved in ethics. Do you mean in what you might call formalized, philosophical ethics, or do you think that actually the term

---

[11] Professor Peter Harper wrote: 'Clarke Fraser, still active at age 97, has confirmed this for me in an email, November 2014, and refers me to his 1961 paper, "What it means to be a medical geneticist", Fraser (1961).' Note on draft transcript, 1 December 2014.

[12] See biographies in Christie and Tansey (2003) – the publication of a previous Witness Seminar to which Professor Paul Polani contributed. For John Fraser Roberts, see biography on page 85.

Figure 3: Professor Bernadette Modell

ethics has changed a bit over the years, and if they were sitting here now they would think they were as involved in ethics as we would say we are at the moment?

Harper: I should think they would be very much involved, and indeed in their later years some of them were. Some like Paul Polani and others. So probably, yes, I think I was thinking of formal ethics. I perhaps would feel that they strongly subscribed to the Hippocratic Oath and that was the foundation of their practice.

Lucassen: Does anybody want to add to that?

Professor Bernadette Modell: I wonder whether part of what you're saying is that they founded the field with a good basis in humanity. It strikes me looking back and remembering my own experiences as a paediatrician in the 1960s that there was a major, major ethical issue at that time, which was whether you did your best to encourage affected children to survive, or whether you thought it was a good thing that they died. I feel that there was an evolution of social attitudes around in favour of doing your best to look after and give equal quality of care to affected children. But I'm quite sure that that was not the general view in the 1960s and there were certainly some areas where it wasn't in my own experience as a young house physician. So when you think about

Figure 4: Professor Angus Clarke, Professor Marcus Pembrey

these founding fathers, can you think of any steps they made, or any explicit statement that one should care for affected people, or was it just implicit in the way they practised? I'm sure that was the case with Penrose because I remember when he came to give a talk in Cambridge and I was a student there. The whole talk was full of humanity, and it gave me a completely different perspective on Down's syndrome.[13]

Harper: All I'd say to that, Bernadette, is that perhaps I'm not the best person to answer because I came into medical genetics from adult medicine, and the people I worked with, like Cyril Clarke and Victor McKusick, were essentially adult physicians where this didn't arise in the same way. So I think others would probably be able to answer better.

Professor Marcus Pembrey: I take issue a little bit with the distinction between people who did the best they could, Hippocratic Oath, and so on, but not embracing formal ethics. I still struggle with quite what you mean by that. But I think the main reason why it wasn't dominant, and that we didn't see chapters in Fraser Roberts' textbook, for example, on ethical issues, was there was so little they could do.[14] So there wasn't the sharpness of the decision and, indeed,

---

[13] See, for example, Penrose and Smith (1966).

[14] Fraser Roberts (1940).

I took the precaution of quickly reading the first edition I was involved in, 1978, just to make sure that there were certain things in there. Indeed, there is a rather discursive short paragraph saying, if I paraphrase, 'Of course, the final decision of what the couple decide to do, whether to take the risk of having a further child or a child, or even getting married, should be left to them.'[15] That was there in the late 1970s, but there was a lot of emphasis on putting things in perspective, and there was an element of directiveness in the sense that it was important to clarify for them what was the background chance of any baby having a severe abnormality manifest at birth, or soon after, and then to put their own particular risk in relation to that.

Certainly, John Fraser Roberts would classify them into low risk where you should be encouraging, and the high risks where they would have to make a serious decision about it.[16] So I think it's partly that there was very little they could do. He even used phrases like, 'It's all rather fortunate that the less we know about the condition and its genetic underpinning the lower the risk, so that makes genetic counselling much easier.'

Lucassen: Can I ask, what would you think then were the first group of conditions where you could do something about it that maybe raised different ethical issues, or began to raise ethical issues more?

Pembrey: Well, I think the biggest one was prenatal diagnosis and selective termination of pregnancy. By the 1970s, of course, we had screening for neural tube defects and so on, and that's when it became crystallized that one had to make very sure that it was the couple aided to make the decision that's right for them rather than anything else.[17]

Professor Peter Turnpenny: Just a small contribution on that point from my friendship with Professor Alan Emery over the last years. He did relate to me the story of finding that in neural tube defects alpha-fetoprotein was very high, and how they found it in the Edinburgh laboratories with David Brock. They came across the discovery and then asked the question, 'Oh, what on earth do we do with this now?', so I think it's very much a case as we, I think, find all the way through, that actually it's the science that leads the ethics. It's the discoveries that ultimately then pose the questions. I think the same is true today. We can

---

[15] Fraser Roberts and Pembrey (1978), see page 292.

[16] See biography, page 85.

[17] For a review of screening for neural tube defects in England and Wales, see Cuckle (1994).

Figure 5: Professor Peter Turnpenny

anticipate a little bit more today, with a very developed discipline of medical ethics, but I think it will always remain the case, probably, that you don't quite know what's around the corner until the science actually poses the questions for you.

Professor Martin Richards: I just wanted to recount something about Paul Polani, which I'm having trouble in dating, but I think this must be the late 1960s, perhaps early 1970s.

At that time there were some organizations, and I believe it was actually the Ciba Foundation that sponsored a series of public discussions aimed at medical students around ethical issues, and it was a kind of travelling roadshow that I was also involved in and went to, I don't know, three or four different medical schools.[18] Paul Polani took part in all this, and he would have been talking about prenatal screening and decision-making around that. I mean not formal ethics, but the point of those meetings was to raise the consciousness of medical students about ethical issues in medical practice, broadly speaking.

---

[18] Professor Martin Richards wrote, 'I don't think this is right; possibly the sponsor of those discussions/ debates with medical students was the Nuffield Foundation. ... [It is] important to note that the range of ethical/social issues in medicine were raised at these meetings were very wide and not just in relation to genetics, prenatal screening, etc.' Email to Ms Emma Jones, 16 January 2015.

Figure 6: Professor Martin Richards

Tansey: May I ask, does anyone else remember those Ciba meetings? It would be very interesting to know how they were started, who got those going?

Richards: They certainly had one in Bristol if that's any help.

Tansey: I'm thinking also of, say, the London Medical Group and the work the London Medical Group did on ethics.[19]

Richards: That was the same kind of activity, yes exactly.

Professor Shirley Hodgson: I was doing a little bit of homework last night, just like Marcus, because I was wondering what to discuss. So I went back to have a look at, for instance, things like the foreword to Lionel Penrose's book on mental deficiency, 'defect'.[20] It was quite interesting that Haldane was saying, even in the late 1940s, that we should encourage people who had mental defects not to have children, and asking whether we would want to offer to sterilize them, and things of that sort, which we would find totally abhorrent now.[21]

I think it's very easy to forget what the climate was in those days. There was an awful lot of feeling that we should not allow people who were not so intelligent to produce more children, and they were worrying that if they had more children

---

[19]   For the London Medical Group, see Reynolds and Tansey (2007); pages 11–15 and Appendices 1–3.

[20]   Penrose (1949); preface by J B S Haldane. Lionel Penrose was Professor Shirley Hodgson's father. See also note 5.

[21]   For a discussion of Haldane's views on politics, philosophy, and science between 1922 and 1937, see Sarkar (1992).

Figure 7: Professor Bernadette Modell, Professor Shirley Hodgson

than more intelligent people, it would result in a drop of IQ in the population. In fact, Lionel, I know, spent a lot of time disproving that, and that there wasn't a steady decline in the IQ of the population.[22] So there was a huge climate of eugenics early on, and this was quite difficult to emerge from. I think that shouldn't be underestimated: various countries were still sterilizing people who had mental defects and other problems, trying to prevent them from having offspring, so that had to be counteracted.[23] I think there were the issues of trying to develop respect for people who did have 'mental defect', as they called it then.

I know Lionel did do some clinics at the Galton, and offered genetic counselling and advice on a small scale. Lionel was saying, quoting Kevles in his wonderful book *In the Name of Eugenics*, that 'a large number of the patients who sought genetic advice acted in a way that would be considered generally to be reasonable so they avoid risks which are serious and accept those which are only moderate'.[24] He predicted that the results of skilful counselling over a long period of years would undoubtedly be to diminish very slightly, but progressively, the amount of

---

[22] Penrose (1950).

[23] For a discussion of eugenics in the Scandinavian countries, see, for example, Roll-Hansen (1989), and Anon (1999).

[24] Kevles (1985), page 258.

Figure 8: Dr Alan Fryer

severe hereditary diseases in the population, based on their own decisions rather than doctors telling people what to do. So I think there was a gradual move from telling people what to do to allowing them to make their own decisions, which didn't happen overnight: this was all happening in the 1950s to 1970s probably.

Dr Alan Fryer: Can I just ask Peter and my more senior colleagues a question about the great men of the past that Peter's talked about. You've talked, Peter, about that, perhaps in their clinical practice, ethics may not have had a formal place. What about in research ethics? When I think of Victor McKusick and many others, there must have been some framework of research ethics in those days, even if not in clinical ethical practice.

For example, all of Victor McKusick's studies amongst the Amish and so on, there are the issues of how you approach people, and consent, and showing respect for the autonomy of people. These are strong ethical issues that actually permeate our specialty today, and have done certainly throughout my time in the specialty.[25] I'd be interested in your comments.

Harper: Well, I don't think there was anything very fixed. I certainly can't remember being aware of any research ethics committees, certainly not while I was in Liverpool at the end of the 1960s, or in America in the beginning of

---

[25] McKusick (1978).

the 1970s. I think it was a much less tangible thing than that. Just to give an example, a Liverpool example for you, Alan: Cyril Clarke's Rhesus study. I seem to remember him referring to this in a slightly teasing way, comparing it with an American study, which was going on at the same time,[26] and saying that the Liverpool study on preventing Rhesus immunization, which was the first series, was carried out on volunteers who were local Liverpool policemen who all volunteered. I'm sure they didn't have to sign anything in writing or anything but they volunteered. He would contrast this, I remember, with the American study, which was done on prisoners.[27] [Laughter] He didn't generalize from that, but, you know, that perhaps tells you something.

Then, as far as Victor McKusick is concerned, of course, the clinical fellows working with Victor, of whom I was one at the end of the 1960s/beginning of the 1970s, we spent much of our time going out into the outback, and it really was outback, tracking down families with whatever. That was an extraordinary learning experience in terms of rural America, and the deprivation and poverty one encountered, but also of the welcoming nature and essentially, well, good nature that people showed, even when it was very clear, and one always tried to make it clear, one didn't have much to offer. But it's fair to say that in exchange what we did offer in these home visits, which took up days and weeks, was just to listen to people and hear about all their medical problems, and occasionally one was able to put them in touch with people who could be helpful. Victor himself was always very keen on making sure that the families got good treatment, which was pretty necessary for things like congenital heart surgery and other things, which they would never have been able to afford otherwise. He also worked very closely with the social scientists; in the case of the Amish, John Hostetler, who was of Amish extraction and had written a book on the customs of the Amish.[28] Victor and all those people working on the Amish with him very much tried to keep in line with the customs and patterns of life of the Amish people, with the result that they were very welcome. The same with skeletal dysplasias – he was always very welcome, as were the fellows, at meetings of Little People of America.[29] Because one tried to provide a bit of help as a *quid pro quo*. You couldn't do very much but you did what you could, and the very

---

[26] Clarke (1968).

[27] See Zallen, Christie, and Tansey (eds) (2004), page 32.

[28] Hostetler (1980).

[29] McKusick, Kelly, and Dorst (1973).

Figure 9: Dr Nick Dennis

fact that you were often going out to see people rather than expecting them to come to see you was appreciated. I don't think there was anything more formal than that, nothing that I can remember.

Dr Nick Dennis: From 1972 till 1976 I was Cedric Carter's trainee at Great Ormond Street, which meant that I sat in on pretty well all the clinics.[30] He didn't allow me to actually do any clinics [laughter], but I spent a lot of time watching him do them. For him, it's worth pointing out that he was primarily a researcher and it was the Medical Research Council (MRC) Genetics Research Unit. I didn't have any sort of clinical appointment, I was an honorary registrar, I was also MRC-employed during that time, so for Cedric his clinical work was one afternoon a week plus ward referrals. He saw his job as giving an opinion and relaying that opinion back to the GP who would then negotiate, if necessary, with the family.

He did have a marvellous woman called Kathleen Evans who was not medical by training, she was trained as a social worker, who sat in on the clinics. That was the beginning of what I think of, I suppose – I hope it's not interpreted as patronizing – as a sort of feminine influence coming into clinical genetics. She

---

[30]   Professor Cedric Carter (1917–1984) was Director of the MRC Genetics Unit from 1964 to 1982. He founded the UK Clinical Genetics Society in 1972. See Wolstenholme (ed.) (1989).

was one of the very early non-medical genetic counsellors.[31] Consultations took about 20 minutes. If somebody required any further discussion because they were distressed, Kath Evans would go out and talk to them in the corridor and might very occasionally arrange to talk to them again later. I should just comment on, well, it seemed, during all my time with Cedric he was very firmly non-directive, and I think I would agree that although ethics was not talked about he had a strong personal ethic of respect for the individual and individual autonomy.[32]

I think it's worth just saying that people didn't talk as much about feelings in those days, and it was a masculine specialty, as most specialties were, and so people kept their views about feelings pretty much to themselves. But I'm sure Cedric really believed that you shouldn't tell people what to do, and I think he saw his role more as encouraging people who had a low risk, rather than trying to stop people with a high risk reproducing. As far as his eugenics was concerned, he did believe in eugenics, but he thought that if people were given information you wouldn't need to tell them what to do because they would make sensible decisions. His eugenics was in relation to intelligence. He thought people with low intelligence only had a lot of children because they hadn't got access to contraception, so he thought that if contraception was available that would take care of itself.

Lucassen: They would be intelligent enough to use it, you mean?

Dennis: Exactly, well, yes, I mean it's a simple decision. And, as many of you know Cedric, had, I think, six children.

Pembrey: Seven!

Dennis: It's rumoured that his main criterion for selecting his wife was on the basis of intelligence. He did one day say to me, 'How many children have you got, Nick?' And I said, 'Only one.' And he thought a bit and said, 'I think you

---

[31] Mrs Kathleen Evans was an almoner at Great Ormond Street Hospital for Children in the 1950s who became involved with family counselling at Cedric Carter's genetics clinic; see Harper, Reynolds, and Tansey (2010); pages 28–30.

[32] Professor Martin Richards wrote: 'There was a lot of discussion of C O Carter's clinic. I don't think there is any evidence that his practice was very different from anyone else in that period. He was a member of the Eugenics Society, as were quite a number of others in the field in that period. He gave advice to couples (wasn't non-directive), including on reproductive matters. In the published follow up study, the authors assess the extent to which that advice was taken. Other follow up studies of other clinics throughout what we might call the pre non-directive period show the same approach.' Email to Ms Emma Jones, 16 January 2015. See also note 33.

should have another', which I took as a compliment. [Laughter] The question of whether people took any notice of the risks in the clinic was to some extent answered by the famous follow-up study of Fraser Roberts and Carter, 'Genetic clinic: A follow-up', where it did turn out that people given low risks had more children than people given high risks.[33]

Pembrey: I'm very glad Nick's told that about Cedric; that's exactly I think what I would say. In terms of getting a handle on this shift fully away from a somewhat eugenic background, it was interesting that Cedric – this was in 1979 but he'd had it in earlier books – felt that the long-term aim of genetic counselling was to see that as few children as possible are born with serious genetically determined or part genetically determined handicaps. He saw the goal quite explicitly as reducing the birth prevalence of these disorders, not necessarily by termination of pregnancy and prenatal diagnosis but restraint or whatever. I think he genuinely believed that if you gave them the information that consequence would follow. Things changed fairly soon after that with the WHO (World Health Organization) having a general statement, which was adapted a little bit for medical genetics to the goal of being to help those families with genetic disadvantage live and reproduce as normally as possible.[34] There was a lot of discussion, of course, about what was normal but it was the shift from the reduction in the birth incidence to that reduction being a consequence of genetic services.

Coming back to John Fraser Roberts, and it was in line with what Cedric thought, he genuinely believed that if you could give people as accurate information as possible and also to disabuse them of stories that they'd heard from relatives and so on, they'd often come with news, even members of the family, saying, 'You've had two affected children, now you're bound to go on having them', and so on. He felt that what I would call therapeutic understanding was part of what they delivered, and I know that John Fraser Roberts did hang up his stethoscope and he put that in the context of not making diagnoses himself. He always got that done by a specialist, if possible, but he certainly believed within the clinical practice of clinical genetics in this therapeutic understanding, as it were. He believed that it was a real benefit in the broadest well-being sense to give them all the information in order for people to make their own decisions.

---

[33] Carter *et al.* (1971).

[34] World Health Organization (1985); see pages 25–8 for discussion of 'ethical issues' in genetics services.

Dennis: There was something I'd forgotten, just one further brief point about Cedric Carter. Yes, in the early years when people would say, 'What would you do, Doctor?', he would never say. But in his later years, he felt that too many people were deterred by what were low risks, and if people said, 'What would you do, Doctor?', and it was a low risk he was prepared to say, 'In your place, I'd probably take the risk.'

Lucassen: I'm really interested in what some of you have been saying about there being no 'formal' ethics then. I wonder if someone can expand a bit on that and what do you mean by formal ethics, and when do you think formal ethics did come in? It sounds like, with what you've been describing about the founding fathers in genetics, that they had very strong ethical principles, so what is it then that is 'formal' ethics?

Harper: Well, I won't answer that, Anneke, or try to, but basically most of us going in as clinicians and coming out as clinical geneticists had absolutely zero grounding in philosophy, psychology, or any other kind of 'ology' really. What we knew we had either picked up through our experience in meeting patients and families, or the general principles we'd been brought up with. We'd no idea that all this had some counterpart in philosophy, or ethics, or humanities, and that's something that I hope we'll come to later on.[35] I remember being really surprised and pleased when I was told, 'Oh well, this is an example of whatever principle', to learn that it actually had a name. [Laughter] I was a bit like a family who had been without a diagnosis for decades and suddenly realising, 'Well, what one's been doing all these years, it isn't just some way-out unskilled thing, it's actually what seems to be a fairly appropriate practice and it's got a name, which is respected among the humanities.' I think that bucked up quite a few of us, learning that, yes.

Turnpenny: In the early 1980s, when the Childress and Beauchamp framework of medical ethics was published, namely the four principles of beneficence, non-maleficence, autonomy, and justice, which we quote an awful lot, and that came out close to the time that I graduated, so it didn't really hit my consciousness at the time, that's for sure, but we often talk about it and quote it and so on.[36] Yet we know it's not particularly helpful for aspects of genetics, especially the autonomy issues, but I just wonder if those who would have been

---

[35] For a Witness Seminar on the influence of the medical humanities in medical education, including medical ethics, see Jones and Tansey (eds) (2015).

[36] See Beauchamp and Childress (1979).

more aware of it being introduced at the time, whether that was a stimulus to think about its application to medical genetics.

Modell: Looking back, I want to ask a question. Of course the concept, the word 'ethics' wasn't commonly used in the medical framework when I was a young doctor, but the first time it came above the horizon for me was with the organization of research ethics committees and then suddenly the whole issue became important to anybody who was interested in research.[37] I just wonder whether the formalization of the ethical principles arose from that. The second point I want to make following on from yours, was the work of Fletcher and Berg and Tranøy. They did a survey of practising clinical geneticists, asking a series of targeted questions to elicit their ethical values.[38] That must have been late 1970s. The three key principles within medical genetics of autonomy, full information, and confidentiality emerged and were so crisply presented that they were useful guiding principles.

Lucassen: My impression is that we are confusing, or overlapping, different types of ethics: what is formal 'ethics' and what are neatly articulated 'principles'? Research ethics appears to be rife with neatly articulated governance principles but what we're trying to get at here is more how ethics entered clinical practice. I don't see that as formal or informal really, so are we using different terminologies?

Richards: Can I bring in a slightly off-side comment? Something I've been interested in writing about recently is the governance of assisted reproduction.[39] I've been reading a lot of material in discussions about infertility, about using donor sperm and all of those things, and what you do see in those discussions is in the, I suppose, late 1960s if you look at let's say the discussion at the Ciba Foundation of those issues that took place, there's a single person who always turns up, who is Gordon Dunstan.[40] He seems to me, he was a kind of one-man ethics person.

---

[37] See, for example, Emanuel *et al.* (eds) (2008).

[38] Fletcher, Berg, and Tranøy (1985).

[39] Richards (2016).

[40] Revd Professor Gordon Dunstan (1917–2004) was Secretary of the Church of England Council for Social Work from 1955 and a minor canon at St George's Chapel, Windsor, and then at Westminster Abbey, London. He was secretary to the group that prepared a report for the 1958 Lambeth Conference, which led to the acceptance of contraception. He was the first holder of the F D Maurice Chair of Moral and Social Theology, King's College, London (1967–1982), later Emeritus, and Chaplain to the Queen (1967–1987). For an obituary, see Shotter (2004).

Figure 10: Mrs Elizabeth Mumford

Lucassen: There's a lot of nodding around the room, yes.

Richards: And I think he, through those discussions, he was always there and he would always bring up the ethical issues, that's what he was invited to do. I mean that was his role in those groups.

Lucassen: But who invited him to do that?

Richards: Everyone. If you look at the formal membership, for example, of the British Medical Association committee, he's everywhere, always the same person. I and others who were around then, I guess the first thing we ever learnt about ethics was listening to him.

Modell: Did he found the first course in King's College on medical ethics? Is that why he was so recognized, so influential? Do you know when?

Lucassen: Elizabeth, behind you, I think might know. It looks like she knows.

Mrs Elizabeth Mumford: I taught on the first course on medical ethics. Ian Kennedy was the director of that course and he is probably an interesting person to bring in at this point. Ian Kennedy was at that time a professor of law at King's College, London. He gave the Reith Lectures on the 'Unmasking of Medicine' in 1980.[41] My time at King's College began in 1984, and that was

---

[41] The six Reith Lectures given by Professor Ian Kennedy can be heard or downloaded at www.bbc.co.uk/programmes/p00h2dg1 (accessed 4 June 2015).

the year Ian Kennedy started, what I believe was the first postgraduate course in medical law and ethics at King's, bringing together people who were graduates in medicine, law, philosophy, and all sorts of other areas to study these issues.[42]

Reverend Professor Gordon Dunstan had retired by that time, but he had been very much involved in the development of the Centre of Medical Law and Ethics at King's, which had been founded a few years earlier, and he often came back to visit. That's certainly how I met him, but he didn't actually teach on the course.[43] I think it's interesting to consider that, while there were interesting things developing in medical ethics in the late 1970s and early 1980s, there were also significant developments at the same time in medical law. I think that was no coincidence; the two came together. The first textbook in medical law and ethics that I remember using was Mason and McCall Smith – the first edition was published in 1983 – the now-famous novelist Alexander McCall Smith.[44]

Pembrey: Yes, I absolutely want to concur about Gordon Dunstan, certainly in the London scene anyway, being the person who was always there on hand. He was Professor of Moral and Social Theology at King's, and I sat on the same ethics advisory committee of the British Paediatric Association, before it was a college, from 1988 to 1995. He was also on that committee and was a huge guidance. I remember, particularly, I asked him what was his personal guiding moral compass – as we would say these days, we didn't use that phrase in those days – and he said, 'principled pragmatism', and 'you go down to the clinic'. I can hear him saying it: 'You go down to clinic with your principles, but you're probably going to have to make a pragmatic adjustment.' That's why I put it in the heading in a chapter of a book I wrote, because it was in honour of this approach.[45]

---

[42] For the postgraduate course at King's, see www.kcl.ac.uk/law/research/centres/medlawethics/index.aspx (accessed 4 June 2015).

[43] Mrs Elizabeth Mumford wrote: 'When the postgraduate diploma (later MA) course began, Ian Kennedy gave the law lectures. I was the tutor. The ethics lectures were taken first by the Oxford philosopher, Dr Michael Lockwood, then later by Dr Raanan Gillon, who was by then the editor of the *Journal of Medical Ethics*. Both of them were writing books on medical ethics at the time.' Note on draft transcript, 12 January 2015. Lockwood (ed.) (1985).

[44] Mrs Elizabeth Mumford wrote, '[He] was originally a well-respected academic lawyer.' Note on draft transcript, 14 January 2014. Mason and McCall Smith (1983).

[45] Pembrey and Anionwu (1996), pages 641–53.

Figure 11: Professor Bobbie Farsides

The other thing I particularly remember about Gordon Dunstan was that he said: 'The key thing is don't get muddled between bad practice and ethical issues; bad practice, if everybody agrees, is just that some are not sticking to what is agreed. Real ethical dilemmas are two opposing or conflicting views, both of which are justified and you have the tension between them.' And then he always followed it up by saying: 'So long as that tension remains, we are safe and probably morally attuned, as it were. It's when there isn't any tension that one should be alarmed about these things', and those are the two things I particularly remember about him; wonderful man.

Richards: He, I know, was a fellow of at least two of the royal colleges, so I mean he was sort of incorporated into the medical world in that sense.

Lucassen: I wonder if someone like Nina, Michael, or Bobbie wants to comment on this discussion about clinicians not engaging in formal ethics and pragmatism and things like that? What do you think listening to this discussion?

Professor Bobbie Farsides: I was very interested by what you said about the roots of non-directiveness lying in a confidence that giving the information in and of itself would lead people to make good decisions, because that's rather a different set of roots for the practice than I might have assumed from the outside. I wonder if things had to become a little bit more challenging when you started to face choices about not just whether or not to have children within an affected family, but also about the termination of pregnancy.

Figure 12: Professor Michael Parker

Then non-directiveness becomes a way of almost putting a barrier around a different moral decision and leaving that with the individuals, whereas, initially, if you felt that giving the clinical facts about the risks involved was the essential information that people needed to make decisions, maybe clinicians felt safer doing that.

Lucassen: So you think the distinguishing thing would be what could be done about those two? Have I got that right?

Farsides: Yes, and why non-directiveness then became such a fixed principle, because it was actually marking out different territory. It wasn't about explaining risk, it was actually about getting involved in moral decisions about termination of pregnancy.

Lucassen: That's very interesting. Of course, as genetic practice has gone on, there have been more and more things that we can do something about, so that would explain that shift to some extent.

Professor Michael Parker: Because it's a Witness Seminar, I don't want to talk about things I haven't witnessed myself but just to report a story. I think it's quite important that many of the figures in medical ethics, Raanan Gillon, Tony Hope, Alistair Campbell, and so on, when they've spoken to me about it, they've talked about the London Medical Group, the student medical ethics groups in

London, which spread to Oxford, as being very foundational for them.[46] Many of them were doctors. Leading to the introduction of medical ethics in medical training, I know there was a campaign by that group of people, which led up to *Tomorrow's Doctors*, the GMC (General Medical Council) report in 1992 or 1993, which said that ethics and communication skills had to be core elements of every course of every medical school.[47]

So there is clearly a story to be told there. I mean, it seems obvious that ethics in some form was taught and has been part of medical practice for a long time, but it was around about that sort of time that it started to be a formal requirement across the UK; that is my understanding at least.

Hodgson: I agree. I think it's very interesting that the idea of ethical discussions being part of the way you dealt with patients is terribly important, but I still feel there was also this climate of telling people what to do and trying to improve the race, which was still going on in the early 1950s, and certainly the 1940s, that people had to escape from.[48] There was an evolution, I guess, from instructing people about how they ought to behave to allowing them to make their own decisions, and it must have been quite a release, the feeling that you could allow people to decide for themselves. Non-directiveness was then the right way forward, which probably came with the ethics principles.

Dennis: I've just thought of something, which was a book by Pappworth called *Human Guinea Pigs,* and that for me was influential – I qualified in 1968 and I think we had no discussion of ethics in my medical training at all – I think this came out about the time I was working for Membership. Have other people heard of this book?[49] Yes.

---

[46] Professor Raanan Gillon (b. 1941) was a general medical practitioner, philosopher, Director of the Imperial College Health Centre, Editor of the *Journal of Medical Ethics* (1980–2001), and author of the book *Philosophical Medical Ethics*; see Gillon (1985). Professor Tony Hope (b. 1951) is a psychiatrist who led the Oxford Practice Skills Course, and was Director of Ethox, Oxford, from 1999 to 2005; see Hope, Fulford, and Yates (1996). Professor Alistair Campbell (b. 1938) was Associate Chaplain to the University of Edinburgh (1964–1969) and Lecturer in Ethics at the Royal College of Nursing (1966–1972), Scotland, when he wrote the first modern textbook in medical ethics; see Campbell (1972). Professors Gillon and Campbell contributed to a previous Witness Seminar on the history of medical ethics education in Britain, and have biographies in the published volume, as does Professor Hope, whose work was also discussed; see Reynolds and Tansey (eds) (2007). See also note 43.

[47] General Medical Council, Education Committee (1993).

[48] See Hanson (2013).

[49] Pappworth (1967).

Pembrey: It was to do with the Hammersmith, wasn't it?

Dennis: It was about research ethics basically, and I think that raised the awareness of ethics in people like me, young doctors, enormously. The other thing that did that in a slightly different way was a book, which was given to me by a fellow medical student who ended up as a professor of medical sociology, so that probably indicates his leaning at the time, and it was called *The Doctor, his Patient and the Illness,* and I've forgotten what the name of the author was; it was by Michael Balint.[50] That just opened up another way of thinking about the medical consultation, which you wouldn't have got from, you know, going to Thomas's and doing your clinical training there.[51]

Lucassen: I think that leads quite nicely into a discussion about presymptomatic and predictive genetic testing and the issues that that brings up, and perhaps how it is different from other clinical practices. We're fast-forwarding quite a few years on from the earlier discussion, when genetic tests became available that predicted something in the future. So Huntington's, I think, would have been one of the first examples of a predictive genetic test, in the early 1990s.[52] I suppose, before that there was linkage for quite a few years, but then actual direct tests to predict something that's going to happen in 10 to 20 years' time came in. That is, I think it still today remains, a core feature of clinical genetic practice. It's quite different to other clinical practices, and I don't know if people want to talk about that and the ethical issues that that raises.

Harper: I was rather hoping that one or two people might expand a bit on this non-directiveness and how did it come in because again this was something that wasn't there particularly at the beginning.[53] Then it became an absolute tenet.

Lucassen: And then it faded again.

Harper: Well, it has not so much faded but people have realised that it's not something that's set in stone.

Lucassen: Yes. And that's quite interesting to relate also to the predictive testing because it comes in very much for Huntington's, doesn't it?

---

[50] Balint (1957).

[51] St Thomas's Hospital Medical School.

[52] The gene for Huntington's disease was identified in 1993; see Huntington's Disease Collaborative Research Group (1993).

[53] Clarke (1997a).

Figure 13: Professor Angus Clarke

Harper: It does.

Lucassen: It moves away again for the conditions where there's perhaps more that can be done. So, yes, I think that's very interesting.

Professor Angus Clarke: By the time I was in medical genetics I feel that non-directiveness was fairly well established in the clinical genetics setting. What I found distressing, and why I cathartically put stuff down on paper was because of antenatal screening and what one could see going on there, where people were put into a context where it was very difficult to say no to what they were being led to comply with.[54]

Lucassen: Can you give an example of what you mean?

Clarke: Well, people booking in for pregnancy and it just being the automatic next thing to do would be to have serum screening or ultrasound or whatever. There's less of it now, but arguably there's still some of that, and I think it's very difficult for midwives booking in pregnant women to have a deep and meaningful, and sustained, conversation with every woman about all the various choices they are going to be making many times a day with new people they don't know. I think it's very natural that people step back a bit from that, and, unless you do, I think the structure of that sort of clinic leads people to go along with the next thing. I feel that's quite separate from what goes on within clinical genetics and the predictive testing and so on.

---

[54] See, for example, Clarke (1997b).

Figure 14: Dr Nina Hallowell

Lucassen: Sorry, I'm probably being a little bit slow here but are you saying that's an example of directive counselling, when the next step is obvious?

Clarke: Not directive counselling but more directive structure to the healthcare system that people are coming into. I mean, within those settings there will be some staff who maybe reinforce it and others who backpedal a bit and try and give people more choice, so there's staff variation but just the structure of the process, I think, is somewhat directive.

Dr Nina Hallowell: I do feel like a bit of a witness because I'm an outsider and a social scientist, and it seems to me that there's been a very big change in terms of ethics within clinical practice, or non-directiveness in presymptomatic testing. So, for example, I can remember getting into this when I worked with Martin Richards in Bruce Ponder's clinic in Cambridge, his breast and ovarian cancer clinic.[55] At that time, the way in which clinics practised was, actually, everyone

---

[55] Dr Nina Hallowell wrote, 'Sir Bruce Ponder ran a clinic in Cambridge at Addenbrookes Hospital with Dr Charis Eng in the early 1990s, partly to recruit people to studies they were initially undertaking to find *BRCA1* and *BRCA2* mutations. As I understand it Bruce was particularly interested in ovarian cancer. Martin Richards will know more about this – ah he says this below! I joined the Centre for Family Research at Cambridge University in January 1994 on a MRC-funded study which looked at families and genetic disorders; this looked specifically at late onset disorders – breast and ovarian cancer and Neurofibromatosis 1. We became involved in Bruce's clinic as they were particularly interested in the ethical aspects of these consultations, namely how to provide women with information so they could make informed decisions about risk management.' Note on draft transcript, 6 December 2014.

was following the model that had been developed for Huntington's disease. If you look at the way this has developed, it was very, very much 'we must be very non-directive in these clinics, women must be given all the information and they must go away and make up their minds', and now there has actually been a very big swerve to being much more directive about certain things, for example, risk-reducing surgery, chemoprevention, and breast screening.

Genetic testing for breast and ovarian cancer now takes place sometimes when women are diagnosed with cancer, so they may make a decision about adjuvant treatment or contralateral risk-reducing breast surgery, with very little of the kind of original model of the non-directive predictive-testing counselling. So it seems that there has been a very big change in the way that counselling is practised in the case of late onset cancers, and certainly the degree of non-directiveness in those kinds of consultations about cancer genetics these days.[56]

Lucassen: And what would you ascribe that to?

Hallowell: I think it's very much to do with the way in which the technology has changed, it's the way in which actually the people who come to the clinics have very different expectations, they have more knowledge of genetics, and those kinds of ethical issues are actually much more understood by the public, the patient groups themselves.[57] It's become much more of a mainstream occupation, I think.

Lucassen: Would you also think that it's due to the treatments available? I think for Huntington's we'd still have those prolonged sessions.[58]

Hallowell: Absolutely, absolutely, definitely. I think at the very beginning, in cancer genetics consultations everyone was very worried about how they might deal with this group of women. A lot of them were very worried, it was very new, the women themselves didn't know much about it, the press had a very particular take on it, and the clinicians were quite worried about how to

---

[56] See, for example, van Dijk *et al.* (2003).

[57] For the impact of media exposure of cancer genetics on the public, in the case of Angelina Jolie, for example, see Evans *et al.* (2014).

[58] Professor Anneke Lucassen elaborated, 'Whilst I agree that consultations may have become more directive for conditions where there are treatments and interventions available (e.g. inherited breast and bowel cancers), for conditions where this is not the case (e.g. Huntington's disease), non-directive counselling over several sessions is still very much aspired to. So it might be less to do with technology of the testing moving on, and more with the condition in question.' Note on draft transcript, 20 September 2015.

actually deliver the services. They followed the best practice that was available at that time for predictive testing, which was the Huntington's practice, and then they realised that perhaps we can actually develop this service and I think the technology and the preventive measures that women can take have really impacted on that. There has been a lot of impact from everywhere and it has changed what is perceived as perhaps an ethical issue in this sphere.

Farsides: I was going to say much the same as you. I think the two cases are different because of the relevance of information. In the first case it's what will or will not happen to you in the future, but in the second case it has now become part of getting the most effective treatment and that, in a sense, takes the heat off the information and makes it a different sort of transaction, almost to collect and provide that information. I was interested to hear from those who were involved in the early days with the Huntington's testing. I wondered whether it was a surprise that people were so reticent to take the tests, or whether those of you with experience could have predicted that, because I've worked with the Huntington's Disease Association, and have been to meetings with young people from affected families, and there's still an enormous amount of ambivalence about having possession of that information.[59]

Harper: Yes, well, maybe things have come back to Huntington's, and I was thinking before this meeting about how, not so much presymptomatic testing began, but what was the background to it? And one of the things is that the idea has been around an awful long time, and I went back to Julia Bell actually. In 1934 in her monograph on Huntington's disease, as part of the *Treasury of Human Inheritance*, she says: '… the almost continuous anxiety of unaffected members of these families over so long a period must be a great strain and handicap even if they remain free from disquieting symptoms. It is thus of urgent importance that some means should be sought by which immunity of an individual could be predicted early in life'; and she goes on to say: 'Development of the science of genetics may at some future date enable us to obtain information concerning the inherent characteristics in such cases.'[60]

That's 80 years ago, and then I remember going to meetings that were always, and as far as I know may be still, biannual meetings involving the World Federation of Neurology research group on Huntington's and the lay

---

[59] For further details of Professor Farsides' engagement with the Huntington's Disease Association, see Farsides (2011).

[60] Bell (1934–1947); vol. 4, part 1, 'Huntington's Chorea', page 13.

Huntington's societies. I think the fact that they were joint meetings was really helpful, but for many years, and I'm talking now about the 1970s and early 1980s, people were looking ahead: 'What if we could predict and how would we, how should we handle it?' There was a series of semi-clinical approaches and neuroscience approaches suggested, none of which turned out to be valid, but they actually gave an important dry run for prediction when it did become possible. So it was 1983 when the first linkage, using DNA markers, was found and that meant firstly that people had already been thinking about it.[61] When I say people, I mean neurologists and geneticists involved in Huntington's and the lay societies involved with Huntington's. People had been thinking about it and not coming to any conclusions, and it was during that era that a number of surveys were done of family members along the lines 'if there was a test available, would you like to have it?' Those surveys mostly showed well over 50 per cent would have it.[62] It was very interesting when it did become possible, we in Wales and others rapidly found that, in fact, a large number of those people who might have wanted it when it wasn't possible didn't want it when it was possible. That alerted us to some of the other problems. The other thing about the DNA linkage was that it was much closer than anybody had any right to expect, so it meant that practical use could be put to this linkage quite a long time before anyone had expected this, so it was a very good thing that it had been discussed a lot.[63]

Then the other thing, which I think perhaps explains why Huntington's has been very much a paradigm, was that there was already a tradition of pooling data internationally and groups would get together and pool data on this and that, and so when prediction became feasible everybody pooled their experience both on the results and on problems encountered, and that meant that one could get really rather robust data from the sheer numbers involved on what problems were being encountered, because people were expecting serious problems. There was a real fear that people, if they'd had an abnormal test, might go out and commit suicide, and it was only the hard data that showed that that was very rare indeed. So Huntington's, I think, led the way, partly because of the links between professionals and lay groups, partly because of this tradition of pooling the data, and also because I think it showed up most of the problems and difficulties as a sort of worst case scenario. I was saying to some

---

[61] Gusella *et al.* (1983).

[62] Evers-Kiebooms, Cassiman, and van den Berghe (1987).

[63] See, for example, Harper, Morris, and Tyler (1991).

people earlier that we, like the other groups involved, rapidly met a number of difficult ethical and practical problems, and my rather naive reaction to this was, 'Well, we can't get away from these problems, let's make the problems the main aim of the study and then even if they're still problems, at least we'll have interesting data.' People will learn more from the problems, and how we did or didn't cope with them, than they will from the actual results. So by the time familial cancers came on the scene in the 1990s there was a good body of data from Huntington's, even though it wasn't really exactly the same situation, there was a good lot that was either the same or quite similar and you could use it as a starting point.[64]

Richards: That's very interesting, Peter, but I'd like to make a comment about something else that was going on, which was actually, I think, almost completely disconnected from what you've talked about, and that is our experience and getting into work with Bruce Ponder.[65] When he arrived in Cambridge, we'd already begun to be interested in some of these issues around testing in clinics and so on. I got a phone call from Bruce one day because he had set up a clinic simply for the families; he was collecting these families, doing linkage studies, and came to the conclusion he would have to give them, as it were, some kind of clinical support. So he set up this clinic and he basically didn't know quite what to do in the clinic, quite literally, and he went to see Martin Bobrow one day and said, 'Look, I need help'.[66] He, for whatever reason, answered, 'Go and see us guys' – our Centre for Family Research.[67] And we started working with Bruce in that clinic, which led us to extend our work into the field on inherited cancer syndromes.[68] I think what was so striking about it was he was completely outside, if you like, the culture of clinical genetics. He was an oncologist, he

---

[64] For familial cancer susceptibility, for example, in 1991 the familial adenomatous polyposis gene was identified; see Groden *et al*. (1991); in 1994 *BRCA1*, one of the inherited breast and ovarian cancer genes, was identified by Miki *et al*. (1994); in 1993, a gene for hereditary non-polyposis colorectal cancer, *MSH2*, was identified; see Fishel *et al*. (1993) and Leach *et al*. (1993). For further discussion of familial polyposis and colorectal cancer, see, for example, Jones and Tansey (eds) (2013).

[65] From 1989, Professor Sir Bruce Ponder was Director of Cancer Research UK's Cambridge Research Institute, now Emeritus Professor. He was Founding Director of the Cambridge Cancer Centre.

[66] Martin Bobrow was Professor of Medical Genetics at Wolfson College, University of Cambridge from 1995 to 2005, now Emeritus.

[67] The research group focused on clinical genetics and families at the University of Cambridge that Professor Martin Richards founded in 1967 and directed until 2005. See Marteau and Richards (1996).

[68] See, for example, Hallowell *et al*. (1997).

was kind of inventing the rules from scratch, as I think was true for some of the other oncologists who were doing the same thing. They were setting up family cancer clinics for families with inherited cancer syndromes, but this was outside the orbit of clinical genetics. Indeed, at least in the beginning, they were not in contact with clinical genetics, which, looking back, I find rather surprising.

Pembrey: Could I just come back a little bit to the transition to non-directiveness at least being the stance, if I can use that phrase, for clinical genetic practice. My own experience was, I was involved in doing research on sickle cell disease and so on, and my first tour in America was in 1976. Following the words that John Fraser Roberts uses, 'genetic advice', when I was giving my talk and then we got onto counselling and I said, 'genetic advice', there were howls of dissent.[69] So I learnt very rapidly the word 'advice' wasn't right. I pleaded, 'Advice: you can take it or not.' But they didn't buy that. That was my first taste of how strong certainly that group was in America. Of course, there was the whole question of screening for, and the lies about the number of people with sickle cell disease in Congress, and actually they used the figure for the sickle cell trait and so on to get money into the system. Then, when it came to DNA analysis, we were the first to get a close genetic probe for haemophilia A and were offering prenatal diagnosis for that and so on. We wrote the first paper in 1984 and then we wrote up our first clinical experience of using it in the *BMJ* (*British Medical Journal*) in 1985.[70]

Over the ensuing years we built up quite a series of family workshops to allow prenatal diagnosis, and when presenting our series in talks it was with considerable relief that we could report that one couple had gone for prenatal diagnosis but had changed their mind and decided not to have a termination of pregnancy. We always tried to say that, of course, you know you can change your mind at any time, it wasn't an inevitable ratchet, and that was a big change because there were people at the time saying, 'Well, there's no point in having prenatal diagnosis if you're not going to terminate the pregnancy.'

Lucassen: We'd still say that now probably.

Pembrey: But, for me, non-directiveness is to allow them to have the prenatal diagnosis in case they change their mind the other way.

---

[69] For John Fraser Roberts, see biography on page 85.

[70] Harper *et al.* (1984); Winter *et al.* (1985).

Hodgson: Actually, you've touched on a couple of things that are very interesting. Of course, one is the prenatal testing because one could ask whether you're being non-directive to the fetus or non-directive to the parents, because the fetus might want to have a say in the matter as to whether they're told whether they're going to get cancer or something later on, so if the parents don't terminate an affected pregnancy obviously this is an issue. But that's another concern. The other thing is the issue of genetic screening. There was, as you said, the sickle cell screening in America, with the misconception that people who had the trait were actually affected, and people who had been screened were feeling that they had been stigmatized.[71]

Clarke: I just wanted to come back to the predictive testing situations, thinking particularly about the coming of, or the prospect of, therapeutic benefit. Then, in a sense, as soon as there's a whiff of that, that sort of trumps non-directiveness so that medical people feel, 'Well, it's obvious, so now you've got to do this.' The issues have changed over the years. There's now the situation of people who are resisting being tested when it's medically obvious that testing would be helpful for their medical management, and they're maybe not always being given enough time to adjust to their situation and so on. I know that's bringing it more up to date and so maybe it's a red herring from a history point of view, but the issue has changed in a rather interesting way.

Lucassen: One of the things I was thinking as the discussion was going on was that I'm still struck by how often people's initial wishes are different to their more considered wishes. So they come in wanting one thing and then, when you have the opportunity to talk to them about it, they actually decide something else and, of course, Huntington's predictive testing is quite a good example, because still far less people eventually have the test than who initially ask for it. Quite often they come in saying, 'Well, I want a test', but once they know all the pros and cons they say they will not have it. I'm interested to know how that relates to non-directive counselling, the difference between the initial and the considered decision?

Clarke: You can be directive about the process in the sense of saying, 'Well, we won't do the test today.'

Lucassen: Straight away, yes.

Clarke: But, you know, it'll be your decision but we'll set the pace a bit.

---

[71] Markel (1992).

Lucassen: We'll set the pace because we know from experience that you might change your mind.

Clarke: But it's a bit paternalistic, isn't it?

Lucassen: Or maternalistic even.

Clarke: But not about the outcome of the decision.

Lucassen: Yes, yes. I'm interested to know what other people think about that.

Farsides: I was just going to go back again, sorry, to antenatal screening and testing, because in 1999 Priscilla Alderson and I were given the very first Wellcome Trust bioethics grant to look at antenatal screening and testing.[72] We rather grandly called our project 'The New Genetics in the New Millennium'.[73] We thought we would go in and talk to people about how genetic advances were going to change the terrain in terms of antenatal screening and testing and what was this new future that they needed ethically to prepare for. We were slapped down pretty quickly and told that people didn't really want to talk about that, they wanted to talk about what they were doing at the moment, which was routine screening and testing for Down's syndrome because they still felt there were so many unresolved ethical issues within that.[74] And non-directiveness was one of the things that we ended up looking at. I thought one of the most interesting features of that was, while people could see the rationale for not being over paternalistic, and not pushing people along a conveyor belt, they were also aware that at certain points if they stood back and refused to be directive or refused to answer direct questions that could be perceived as being directive – their patients or the women they were working with felt abandoned by them.[75] So there was a real sense of letting people down by adhering to something that had sort of noble intentions but actually might stop you assisting someone in their decision-making when they felt that they needed it most.

---

[72] Priscilla Alderson is a sociologist of childhood, and has worked at the Social Science Research Unit of the Institute of Education, University College London where she is Professor Emerita of Childhood Studies; www.ioe.ac.uk/staff/SSRU_2.html (accessed 24 June 2015).

[73] See Alderson, Williams, and Farsides (2001).

[74] Williams, Alderson, and Farsides (2002a). See also Williams, Alderson, and Farsides (2002b).

[75] Williams, Alderson, and Farsides (2002c). See also Williams, Alderson, and Farsides (2002d); and Williams, Alderson, and Farsides (2002e).

Lucassen: I wonder if we can move on? I think this discussion lends itself quite nicely to thinking a little bit about consent and confidentiality issues raised by genetics. If we go back to Huntington's as an early example of maybe testing somebody at 25 per cent risk when the person at 50 per cent risk doesn't want to know, how do we manage the tensions that raises in consent and confidentiality?

Harper: Well, since you bring up Huntington's and the 25 per cent risk situation, it wasn't something I'd thought would come up today but I do remember it causing a lot of quite agonized discussions in various forums. And the upshot was that two groups, our own in Wales and a group in Leiden, looked over our, by that stage, very large series of several hundred requests for prediction, and what we found was that requests for people at 25 per cent only made up a small proportion and that for most of those there weren't major issues because the intervening parent was no longer living or because they were quite comfortable about things.[76] And so there was the really difficult question of what do you do if a person at 25 per cent risk wants to be tested but the intervening parent doesn't want to or something like that? Both our studies found, I think, either one or two examples of that and almost invariably what was a genuinely very difficult situation could be got round by good counselling and preparation beforehand. Both our groups concluded from that that there wasn't any point issuing grand guidelines about what to do in such a situation because it was a bit like enacting a law for something that is very, very rare and occasional, and it's going to cause more trouble than it solves.

That was an example of how useful these large data sets were in coming to what was a very pragmatic conclusion, which was you should spend most of your energy not getting into that problem in the first place, and that the occasions when you can't get out of it are too rare to try and generalize from and it's probably better to do it on a case-by-case basis.

Lucassen: I suppose I'd thought of consent and confidentiality as a more general issue, of it raising communication within families in a way that, again, other specialties don't do quite so much, although they, of course, also do.

Parker: I'd like to say lots of things about the stuff you just mentioned but I'm quite interested to ask a question. One of the things that Marcus talked about earlier raises the question: what's an ethical problem?[77] Is it only an ethical

---

[76] See, for example, European Community Huntington's Disease Collaborative Study Group (1993).

[77] See page 24.

problem, you were quoting someone else, when people disagree, or can there be an ethical problem when people agree? And one of the things that struck me when I first got into thinking about ethical issues in genetics, which was in the late 1980s and 1990s, was the extent to which there was disagreement. So far we haven't really touched upon that, and it seems to me that there's disagreement between individuals sometimes but also between different regional genetics services.

Also, the point about consent and situations where someone wants a test and there are other people at 25 per cent risk, because I've personally found there to be a lot of disagreement, and certainly historically when I first started. Both sides of this discussion took the view that ultimately you'd try to encourage, initially you'd encourage family members to talk to each other, but the differences emerged when that wasn't possible, when people didn't want to do that. There were some people who felt, 'Well, ultimately the person in front of you is the person you should be taking care of and they should get access to the test, and other people shouldn't have a veto over that.' But there were other people who felt it was really important not to test other family members without their consent, and had worries also about how they might hear about the information and the result of the test. There were those who thought very strongly you shouldn't give the test to the parent. I just wonder whether that's a starting point for a discussion potentially about differences of views about ethics historically, and whether people have had experience of that.

Dennis: Well, I've been retired for seven years now so perhaps I shouldn't comment on the present state, but I would not entirely agree with that because my experience was that there was a big consensus among regional genetics units about the general approach. Of course, we took part in the Huntington's meetings when all the regions got together and pooled their experience and that was very helpful, but I didn't feel there was a lot of disagreement about the basic approach that we should be adopting. I think the big contrast is between clinical genetics and the rest of medicine. Any of us who have done genetics clinics will be familiar with referrals that come with an assumption that we are going to tell this family not to have more children. And we'll be familiar with other clinicians who would say, 'What do you geneticists do?' Then, when you start explaining they're clearly thinking, 'You mean to say you spend all your time doing that?' [Laughter] And how we got into that situation, I think, is linked to the non-medical genetic counsellors coming in and highlighting issues, highlighting things that the largely male workforce up to that point just took as assumptions without making them explicit.

Hodgson: I would develop on that because certainly in cancer genetics, which is my main specialty, we find that the genetic counsellors really come into their own because most families in the end will make sensible decisions. It is the counsellors who are often able to see a person, for instance, who is reticent about telling her own *BRCA* test result to her daughter, possibly because she's feeling guilty, or because she doesn't know the right time to tell the daughter, or feels that the daughter won't understand, and determine what the reasons are beneath behaviour that might appear to others to be intransigent. The genetic counsellor is much more able to dissect out people's fears and motivations regarding the communication of their test results to other members of their family, and perhaps allow people to come to a more open decision that will benefit the other family members than they would otherwise have done if they had just been talking to the busy doctor.[78]

Clarke: What Shirley was just saying reminds me that there's a nice but very intermittent and occasional strand of other disciplines coming out with really helpful insights into family communication. I'm just reminded of a paper from the late 1970s about polycystic kidney disease and about, basically, family cycles of mis- or non-communication, that are self-perpetuating, so you find out about something in the family in a very distressing and difficult way.[79] You therefore clam up about it, don't tell your children, and the same thing happens to the next generation too. So, looking at other medical specialties, I agree there are a lot of problem areas, but there are some really nice beacons of good practice and insight.

Lucassen: Nina, you did a study looking at communication by men of *BRCA1* and *BRCA2* results in families, didn't you?[80] I don't know if you want to say anything about that?

---

[78] Professor Peter Harper discusses the professional specialty of the genetic counsellor in his textbook, *Practical Genetic Counselling*, Harper (1981, seventh edition 2010), subsequently revised by Professor Angus Clarke; see Clarke (2015).

[79] Manjoney and McKegney (1978–1979).

[80] Hallowell *et al.* (2005).

Hallowell: Yes, I would agree with what Angus has just said actually. A lot of the family work, not just my own but lots of work in Cardiff, suggests that there are definite familial patterns of communication about health and other issues. I agree there can be lots of people who can actually add to this, lots of communication studies. People have done some very good work, particularly the group in Cardiff.[81]

Pembrey: Just coming back to Mike Parker's point about disagreement between clinical geneticists; it's relevant, only just, to his point, and that is this question of having a group discussion around the patients that one is going to see in the clinic. This was the importance of being able to have a big enough group that involved genetic counsellors and so forth to discuss these issues, which I think is very important because you start off with perhaps a difference of opinions, but in the end you've got to come to some sort of consensus about what you are going to say in the clinic. This was particularly brought home to me when I was helping, with others, to do some training in Portugal when they were setting up their service there.[82] In the end there were five clinicians and five laboratory people trained, and almost towards the end they said, 'Right, we'll be able to have five centres?' And I said, 'No, no, you can't possibly have five centres, you at least have to have two clinical geneticists in the same centre. You can't, in my view anyway, handle this unless you've got somebody also trained to talk about it.' So I think the structure of the way clinical genetics has developed with, generally, pre-clinic discussions with each of the patients, which is not that common in other parts of medicine, helps to resolve some of these differences.

Lucassen: I would echo what Mike says. A lot of people talking about ethical issues in their clinical practice want to know what the set of rules is that they should use to resolve the issues. If they use different rules they might initially at least think, 'Actually, you can't do that, you've got to think of the patient in front of you', or 'you've got to think of the wider family'. And I think you're absolutely right, that could result in apparent disagreement, but if you talk that

---

[81] Dr Nina Hallowell wrote, 'Srikant Sarangi's group: This group was working as part of Cesagen's work package. Cesagen was the genomics centre that was based at Cardiff and Lancaster universities and was part of the ESRC funded genomics network. This network was set up to undertake social science and socio-ethical research in genomics.' Note on draft transcript, 6 December 2014.

[82] Professor Amandio Tavares from the University of Oporto invited Professor Pembrey to assist with training for the MSc in Medical and Human Genetics from 1985 to 1987. Professor George Anders was involved, but mainly Charles Buys (1942–2014), who was Professor of Human Genetics, University of Groningen, Netherlands. Note on draft transcript from Professor Marcus Pembrey, 27 August 2015.

through then eventually you'll probably end up agreeing much more than at your starting point. Certainly, the stories we hear in Genethics seem to disagree with how things should be managed at first even if more common ground is found later. I think that's what you were hinting at.

Clarke: I'm really interested actually in the persistence of that dispute about the one-in-four risk for people in Huntington's families. Clearly there are centres that don't practise the way I do.

Lucassen: Shocking. [Laughs]

Clarke: Quite, clearly [laughs]. That's despite the best of intentions and a lot of discussion, and Nick might be pleased to know that the Huntington's disease forum – the UK Predictive Testing Consortium – carries on and still functions because, you know, it's not running out of things to say – a bit like the Genethics Club; there are still lots of things to say.

Richards: Could I just make an observation, which is simply the growth in numbers of people on the ground. In the days when we first got involved, it was more or less the case that we knew, face-to-face, people in every genetic centre in Britain, and they came together and they talked with each other. I don't know what the rate of increase has been but when I first knew anything about clinical genetics in Cambridge there was one part-time person who lived in a Portakabin.[83] There's now a tower block. This really goes back to Mike's point that there is now a large group of people doing the same thing; presumably that does lead to consequences and different styles of practice developing and so on, in a way that just didn't seem to be happening in the beginning.

Parker: Yes, I do want to pick up the point. I used that example as an illustration, but I think it's the fact that we've heard about this long history of thinking about ethics, and we're here today, and I think it's more, it seems to me that it's got to be more than simply that these are sensitive topics or that they are difficult to talk about, or that these are politically troubling. Ethics, if it's anything, or at least a certain important element of it, has got to be about the fact that these are problems or situations in which there are different, potentially conflicting, ways of answering the question. It's not just about who should answer the question; I think it's partly about that. It seems to me the story we've just heard

---

[83] Dr Nina Hallowell commented, 'For example, in cancer genetics there was the cancer genetics group, an informal group of people doing clinical, molecular and psychosocial work in cancer genetics, now I think this is a formal grouping within the BSGM (British Society for Genetic Medicine).' Note on draft transcript, 6 December 2014.

is, at least in part, a struggle about how are we going to resolve some of these, not necessarily conflicts between individuals, but certainly tensions between different considerations and concerns. We heard at the very beginning from Peter about some of the virtues, the character of some of the people working in the field and their struggle with these difficult situations. Partly they're struggling because there are different things that are important to them that are pulling in different directions. That's what I was trying to get us to say something about really, rather than actually say anything very specific about…

Lucassen: Disagreements, yes.

Parker: I think we've heard enough about that, but I do think that's important and enduring.

Farsides: I wonder, Martin, if you think the interest that social science started taking in the matter is also helpful in identifying new potential harms, or a different sort of moral consideration, for example, something like stigma, which really is an issue in sickle cell disease, particularly. So there's something about the cross-disciplinary interest in genetics that then feeds into the growth of the list of the morally sensitive issues to consider.

Hallowell: I'll answer for Martin Richards because I've been doing that for 20 years now [laughter].[84] The answer is, yes, he would think that, wouldn't you, Martin?

Richards: Yes.

Hallowell: I just wanted to add something to what Angus was saying, a very quick comment, and we'll talk about it later, which is the addition of the genetic counselling people, because actually the genetic counsellors have changed, and the profession of genetic counsellors has actually changed this field in a huge way. Also, I think the genetic counselling profession is very much focused on ethics and non-directiveness, obviously, but on ethical practice and that is actually at the core of the profession and quite a lot of their education is based on that.

Hodgson: Just a quick word on the genetic counsellors, although we're going to talk about them later. It was interesting when I did a survey of the cancer genetics services in different parts of Europe; an awful lot of them didn't have

---

[84] See biography on page 86.

genetic counsellors or didn't give them any responsibility, and having them as part of a team has made cancer genetics services more deliverable in direct relation to the presence and acceptability of genetic counsellors.[85]

Harper: Anneke, I was just going to mention one thing, which is down on the programme. Nobody's touched on it yet and this is this question of how unique most of these ethical issues are to clinical or medical genetics, or how much are they universal? I've seen in the past that this caused a few dust-ups, mainly people saying, 'Oh, this genetics is nothing very different from everything else.' For what it's worth, my feeling, based on what I've encountered over quite a long time, is that indeed most of the difficult issues that give rise to ethical problems are not unique to medical genetics, but they are certainly an awful lot more prominent. Very often they've been lurking there in other fields of medicine under the surface, or at least unrecognized by the people involved, or maybe they have then started to emerge. So half of the value of all the work that's gone on in the genetics field has, I think, been to alert other people to the fact that, no, these ethical problems aren't absolutely unique, but we seem to meet the sharp end of them. I think, as geneticists, we've done a fairly reasonable job at exposing these issues and arguing about them in a way that has been unusual in other medical specialties.[86] This rather leads on to the challenges for others. Well, if these issues aren't particularly unique and you begin to identify them, how are you going to cope with them? I think that's certainly become true in the cancer field, but also in the neuroscience field. There's a sort of neuroethics that has come along in the wake of ethics in genetic issues.[87] I don't know what other people feel, but my feeling is really the issues aren't that unique, but maybe the degree and complexity to which they affect practice are, if not unique, certainly kind of unusual.

Turnpenny: We face the situation in cross-talk with our colleagues and with managers especially; the whole healthcare system is geared up to the individual, largely, to the person with a symptom who comes and needs to be assessed

---

[85] Hodgson *et al.* (1999).

[86] '...the field [of medical genetics] attracted medical doctors whose background in general medicine and paediatrics had made them aware of the difficult, often tragic, problems faced by families with serious genetics disorders and more able to relate to families in discussing these problems. Having the time to do this through the necessary steps of taking a pedigree and going through the various risks and options was an important factor, one largely lost in more pressured medical specialties', quoted from Harper (2008), page 456.

[87] Illes and Sahakian (2011).

and investigated and treated and so on. I think we're always in those sorts of discussions pointing out that the true patient in the genetics clinic is the family, not only the individual. This we, of course, know, but we see that it's quite difficult for others to fully, fully appreciate what the implications of that are. It affects how we document what we do, it affects the type of notes we have because we have family files, which contain details about many individuals, not just one individual. It determines how we talk about families rather than individuals. It has a big effect on the way we investigate people with genetic testing, of course. I think that's one of the key points in the transition, or at least to understand the unique place of ethical issues in genetics, is the place of the family as opposed to the individual. It's still an issue communicating that and helping people to understand it.

Fryer: I remember doing one of your clinics, Peter (Harper) in Swansea, and being faced with the issue of a child who had a parent with adult polycystic kidney disease, and the issue was that of testing the child by doing an ultrasound. It was not long after I'd started working for you, and I'd come from a paediatric background, and in paediatrics my experience was if the parents requested it, you just did it. Then I came back to Cardiff and we were talking about this issue. Marcus was talking about pre-clinic discussions; we had a lot of post-clinic discussions in Cardiff as well at Thursday lunchtimes, I remember well. And the discussion presented the question : 'Well, is this the right thing to do?' Obviously, we were then very influenced by your experience in Huntington's and so on. So I think Peter's point is very, very well made; the issues may not be different but they were crystallized in genetics, in this case the particular issue of genetic testing in children.[88] This issue was crystallized in clinical genetics practice and then that brought it to the attention of paediatricians and others dealing with children that, to be honest, having been a paediatrician, I hadn't really thought through.

Lucassen: Yes, polycystic kidney screening in children is a perfect example, isn't it, of that bridge between paediatrics and clinical genetics.[89] Angus, I don't know if you want to say something about the history of genetic testing in children?

Clarke: Well, I was going to say something to come back to Peter Turnpenny's remark actually about the family being the patient, and to tie that in with

---

[88] See Clarke (1997c).

[89] Professor Anneke Lucassen wrote, 'In that a kidney scan in a child might reveal an inherited tendency to adult disease.' Note on draft transcript, 16 October 2015.

what Bernadette mentioned about surveys, international surveys – I think you mentioned, didn't you, Dorothy Wertz and Kåre Berg and Fletcher, yes[90] – because, of course, families take different shapes in different communities and societies, so you get systematic differences in practice and attitudes to things like non-directiveness and everything else in different parts of the world. That's a whole issue that we probably aren't going to get on to, and explore, but it's worth flagging up the fact that a lot of people who have trained in one place, say North America or Europe, then go and practise in other parts of the world. This means that you've got professionals who are socialized into one approach to dealing with families then trying to train their colleagues and work with families who are expecting something different; it still creates lots of conflicts and tensions.

Lucassen: Do you also want to say something about genetic testing of children?

Clarke: I suppose that emerged out of the predictive testing for Huntington's work really. I think David Craufurd talked about it, and then Peter Harper and I wrote a paper, and then there was the Clinical Genetics Society working party.[91] There have been several working parties since then so the practice has changed slightly, and I think the principle is laid down of trying to preserve the autonomy of the child for a future where that isn't going to put them at a disadvantage medically or in some other way.[92] That principle is still current. It takes slightly different shapes in Europe maybe from how it does in the UK.

Lucassen: I think that as a principle that can take a bit of time to communicate, so I think, despite all those guidelines, we still get parents saying, 'I want my child tested', but then, when you talk to them about the possible disadvantages of testing they get to an understanding of the reasons for possibly delaying testing that they don't have when they first come in to ask for it. That hasn't really changed over the years, but what has changed is the technology or the ability to test, so it's easier to get the test done without that discussion, I suppose, so I agree with you, the principle is interesting.

---

[90] See page 21, and Wertz and Fletcher (1993).

[91] Dr David Craufurd founded the first multidisciplinary management clinics for Huntington's disease in the UK; www.manchester.ac.uk/research/david.craufurd/research (accessed 5 November 2015); see, for example, Tyler, Ball, and Craufurd (1992). Harper and Clarke (1990).

[92] Borry *et al.* (2009), and European Society of Human Genetics (2009). See also Clarke (1994).

Clarke: Yes, there's probably less weight put on carrier testing now than there was, I suppose the same with newborn screening for sickle cell, at least in England, in identifying all the sickle cell carrier infants.[93] Then, obviously, that was a completely clear conflict between the genetics professional guidelines and the newborn screening practice as it came in, so that caused a degree of unhappiness. That was quite an interesting clash of ethics. Well, in Wales we don't identify haemoglobin, sickle globin carriers, because we use a slightly different technology that deliberately doesn't generate that information. But in England the sickle carriers are all identified. So, in terms of notifying families of results the vast majority of results are carrier results not affected results, which raises workload questions and all sorts of things.

Modell: Yes, and whether one is treating families or individual patients.

Lucassen: But also, a little bit about whether one is setting out to test or whether one has found it as a result of looking for something else. That's another difference, isn't it? So if you find carriers as a result of looking for affected children that's different to testing specifically for carrier status.

Modell: Yes, but then you're making a choice about whether you're case finding or whether you're a geneticist looking for families who may benefit from genetic information.

Lucassen: Absolutely. No, I think the only point I was making is that there is a difference between setting out to look for something and finding it as a by-product of the test.

Dr Mark Bale: It's just an interesting reflection; I'm relatively new to all of this, and I'm finding it fascinating to see what we can do with some of our horizon scanning. Even though I'm relatively new, I'm still one of the people with the longest experience of clinical genetics in the Department of Health (DoH). I got involved in genetics generally in, I suppose, the mid-1990s. In the Department, people come to me assuming that we've got the ready-made solutions to, say, what do we do about CJD (Creutzfeldt–Jakob disease) testing, or what do we do about Alzheimer's testing is now coming up, that they are going to be able to pick up and take off the shelf a kind of model that works for genetics.[94]

---

[93] Sickle cell disease screening for newborns was introduced in England between 2003 and 2006; see Streetly, Latinovic, and Henthorn (2010).

[94] See for example, Bechtel and Geschwind (2013).

Figure 15: Dr Mark Bale

I think it is coming back to what you said; it does seem that genetics has had to take these steps and work on practical issues that now other people are coming into. What tends to happen when they talk to me about what their system is for genetics is they go away scratching their heads, 'How on earth are we going to do this in practice?', and it still happens in screening. Screening people assumes that we will be able to give them a solution on things around carrier testing, how that's going to be recorded and transmitted to families, and we tend to say to them, 'Well, I thought that was your area?' [Laughter] It's quite an interesting one because it just reminds me actually that genetic testing has been in the vanguard for quite a lot of this, and we should learn some useful lessons from that.

Harper: A very quick point while we're on the screening area; it's something I just made a note of on the train coming up. One of the reasons why, probably quite numerous reasons why, geneticists and other clinicians such as obstetricians often have very different views about screening, particularly antenatal and generally reproductive screening, is that geneticists have been brought up in a Bayesian world.[95] On the whole, most of their practice is with high risk families, but they are used to the concept that the person who is at high risk, and worried

---

[95] 'Bayesian analysis is commonly used to calculate genetic risks in complex pedigrees, and to calculate the probability of having or lacking a disease-causing mutation after a negative test result is obtained'; quoted from Ogino and Wilson (2004), page 1.

and comes to see you, is utterly different from the person out there in the population who usually is unaware of the problem, maybe doesn't want to know about a problem, etc. For geneticists, it's always been one of the tenets; these are different situations, radically, and you have to handle them differently, whereas for many others, and I think quite a few epidemiologists and public health people are included in this, it's all lumped together and we're feeling a bit, that what's good in a high-risk situation, if you can do it technologically then you should be able to apply it in a screening situation. Of course, you can't always do that. So, although it's nothing to do with ethics, it actually does underpin quite a few of the ethical differences and dilemmas which occur, especially in the screening field.

Lucassen: Before the break we were talking about genetic counsellors and how they've influenced both the profession in general and the consideration of ethical issues. One thing, of course, to note is that there are no genetic counsellors here today.[96] That's not for want of trying, but unfortunately none whom we approached were available today. From a personal perspective, one of the biggest changes I've seen in clinical genetics is to do with genetic counsellors. When I started in clinical genetics, I always had a genetic counsellor in clinic with me, and co-counselling was a valuable and valued part of the clinical approach. Recent financial cutbacks to NHS (National Health Service) genetic services, and in the face of increased referral rates and thus waiting times, means that we now have to see patients separately.[97] In fact, the last time I had a genetic counsellor with me was probably about a year ago, and that's a loss to the service I can provide. It makes a huge difference to clinical practice: if you've got two people in a room, and a genetic counsellor is there to co-counsel with you and pick up the cues you don't pick up yourself, that's incredibly helpful for the overall delivery of genetic services. I think it's a huge loss that we don't have that anymore. But I wondered if we could use that starting point to go back and say

---

[96] The development of genetic counselling was discussed at a previous Witness Seminar to which the counsellors Mrs Lauren Kerzin-Storrar and Professor Heather Skirton contributed; see their comments and biographies in Harper, Reynolds, and Tansey (eds) (2010); pages 68–73, 124, 129. See also McCarthy Veach, LeRoy, and Bartels (2003), Chapter 2, Overview of genetic counseling: history of the profession and methods of practice, pages 23–37; and Harper (2010), pages 4–5. See also note 110.

[97] Professor Anneke Lucassen wrote, 'These [cutbacks] were gradual, but I noticed the effect on counselling around 2012/2013, and were perhaps more to do with increased referral rates with static funding, but at local NHS Trust levels there are often year on year cutbacks of five to ten per cent. In larger specialities these can often be absorbed but in small specialities like genetics this means fewer staff for the same or increasing patient referral rates.' Email to Ms Emma Jones, 21 September 2015.

a little bit about genetic counsellors coming into the profession? Angus, given that you're in charge of the training programme for genetic counsellors, I don't know whether you want to represent genetic counsellors today?

Clarke: I certainly couldn't do that, I wouldn't be allowed to. I can reflect just a little bit upon their changing nature. They were initially called co-workers, associates, and I know there was a pool of people in that role when I started in Cardiff's genetics service in the 1980s, which consisted of a social worker, a psychologist, a science graduate, a couple of nurses, and a slightly fluctuating body of a few more who shared an office and would always have words of wisdom and a lot of experience coming from very different areas. I found that a very good model, of having people from many different backgrounds working together. Then, what happened is there was a bit of professionalization of genetic counselling, and it became very nurse dominated for a bit, and to me that was less interesting, the fact they came from a very uniform background. I think the role of nurses is really valuable in that group, but I think when it looked like it was becoming almost totally nurse dominated, I felt it lacked something. Now, both Manchester and Cardiff have courses for genetic counsellors, so we've got to a predominantly non-nurse body of genetic counsellors.[98] There are a few nurses and other healthcare professionals in there, but I think nothing like enough now. I guess these things come in cycles, and we'll have to work out a way to make it easier for healthcare professionals, established healthcare professionals, to move through some sort of training into the genetic counsellor role.[99]

Lucassen: The general idea that the genetic counsellor adds to the genetic advice or genetic information that the clinical geneticist gives is something that is worth picking up a bit more because, as I said, historically that's fluctuated quite a bit, hasn't it?

Hodgson: There are two issues here. I think genetic counsellors certainly add a great deal to the genetic counselling clinic's agenda because they come to the clinic with a different background. To go back to my survey, it was quite interesting that in Germany, for instance, there was a lot of antagonism to the

---

[98]   MSc programmes in genetic counselling are available at the University of Manchester; see www.mhs.manchester.ac.uk/postgraduate/programmes/taughtmasters/geneticcounselling/, and Cardiff University, where the course was founded by Professor Angus Clarke; see http://medicine.cf.ac.uk/cancer-genetics/research/medical-genetics/our-teaching-and-training/msc-genetic-counselling/ (websites accessed 25 June 2015).

[99]   See, for example, Skirton *et al.* (1998).

idea of having genetic counsellors, who were not doctors, doing any serious counselling.[100] They've only just started taking on a few counsellors to share some of the clinical responsibility of genetics clinics, particularly cancer genetics. But the other thing is that, actually, genetic counsellors can do a lot of the cancer genetics without doctors, so long as they work as a team. So they've found a role that perhaps they didn't used to have, but they have a lot of training that I certainly didn't get in terms of picking up cues about family emotions and relationships, and they have a very important role.

Lucassen: One of our other headings is the involvement of social scientists in in clinical genetics services (see Table 1).

Hallowell: I just wanted to pick up on this issue around the genetic counsellors and what I said earlier. I have been the external examiner at Manchester, and I'm currently the external examiner at Cardiff for the MSc courses, so these are the two courses in the UK where genetic counsellors are trained. It seems to me that what is absolutely fundamental to these courses is this idea around ethical issues, at every single level. That's particularly clear when you come to look at the student dissertations, which are always about what I would call possibly 'lived' ethical issues rather than your four principles of Beauchamp and Childress.[101] I think that ethics is actually very integral to that training; I'm not suggesting for a minute that the clinicians among you are not ethical, it's just that they focus very much on lived ethical issues: consent, confidentiality, autonomy, are all very much woven into their training at every level.

Clarke: To come back to Shirley's point about Germany: until very recently there were financial reasons why German genetic centres would not want to have counsellors, and that's because they got all their reimbursement by tests performed. They were apprehensive that counsellors would talk people out of having tests [laughter] so they would undermine their own basis for salary. But that has now changed and they can bill for the communication counselling side of it. That's the last few years.

Fryer: Just going back to the history again. I think that while Shirley's comment about variation across Europe is important, actually, when I came into genetics there was quite significant variability across the UK. I trained, as I've said before, with Peter in Cardiff, and perhaps Peter will say something about this later,

---

[100] See note 85.

[101] See page 20.

as to how that ethos of having the involvement of genetic counsellors came about, which, as Angus rightly said, included Audrey Tyler from a social work background. Indeed, there were a variety of different disciplines among the group in Cardiff. When I came to Liverpool, we had what was called a genetic health visitor, and we actually had two genetic health visitors. Although we were a very small department, we were built along the same model as Cardiff and my closest neighbours in Manchester where they, of course, had quite a lot of genetic nurses or whatever they were called at the time and, in addition, they had Lauren Kerzin-Storrar as an American who'd come through their genetic counselling training programme.[102] They became very pro developing non-nursing genetic counsellors in Manchester because Lauren made such an impact on them. They were very pro the development of individuals coming from a genetic science background and then training them as counsellors as part of the team alongside nurses.

Meanwhile, I think there were a lot of other departments across the country which did not have any genetic counsellors or other co-workers. From an historical point of view, it's very important to realise that this was built up in a very piecemeal way. The role of genetic counsellors in the way you've described, and what I experienced in Cardiff and then took into my own practice in Liverpool is not the way that the role of genetic counsellors has developed in other centres – I think that has to be recognized.[103]

Dennis: Well, my personal experience may be worth recounting. I've mentioned that I started with Cedric Carter in Great Ormond Street in 1971/72, and the role of Kath Evans there; she was one of a group of three or four people who were employed as researchers.[104] Cedric had these big studies where he followed up families of children who had been through Great Ormond Street with various malformations, and so he needed a team of people to go out and interview families to find out recurrence risks, and Kath was primarily one of those. She was trained as a social worker, but a very helpful addition to the clinic it seemed to me. Then, in 1976 I went to Buffalo in New York State for two years, and there was a lady called Gillian Ingall who was also a social worker by training, or originally she was one of the ones who had been a medical almoner

---

[102] See note 96.

[103] For a discussion of the profession of genetic counselling and training internationally, see Bowles Biesecker and Marteau (1999).

[104] See note 31.

but she'd gone to America with her husband who was a surgeon, had taken an interest in genetics and got a job.[105] And there was Barbara Bernhardt, also from a science background there, so they both functioned as non-nursing genetic counsellors. Gillian was quite scathing about the people emerging from the genetic counselling degree course in North America, because she felt they didn't have the sort of hands-on experience you got if you had a nursing background.

I came back to Southampton in 1978, and I was trying to think why was I so keen to get somebody to support me in the clinic. I think it was just isolation because, apart from Elspeth Williamson, [a clinical geneticist], who was part time, I felt very much on my own, I felt completely unprotected. I was sitting in the clinic trying to do something that I really didn't feel I was properly trained to do, and I just felt it would be so helpful to have someone who could have a separate viewpoint on some of the things that I was having to discuss with families. So I tried very hard to get somebody, but I tended to think of it as someone from a nursing background and in fact when we got the money a few years later, we managed to appoint three people all from nursing backgrounds. I thought at the time it seemed very helpful to have somebody with a medical insight, and hands-on medical ability to sit with and talk to somebody sympathetically, but also know what the medical problems were. There weren't the courses, degree courses in genetic counselling hadn't started up at that point, so we were very lucky. I mean we ended up with amazingly bright people from a nursing background, which is one very good way of doing it, I would say.

Turnpenny: I've been involved in the Peninsula Clinical Genetics Service for 21 years now and we had a half-time nurse when I started, and then began to appoint some new folks soon after.[106] We appointed from nursing, established nursing backgrounds, health visitor backgrounds, and it seemed to work very well. The one great advantage of such a person was that they had a bedside manner, generally speaking, and for somebody who was a health visitor they had this great ability to get across the doorstep, as I always used to say. A lot of work was done by home visiting at that time, much less now, of course. To win the confidence to get into somebody's home, that was a great asset. I have to admit to being a person who was a little bit worried about people coming straight off degree or MSc courses with their genetics training but no care training, or

---

[105] Gillian Ingall was a President of the National Society of Genetic Counselors in the USA, see Heimler (1997).

[106] A service for the population of Devon and Cornwall, in the south-west region of the UK; see www.peninsula-genetics-service.org/index.html (accessed 25 June 2015).

healthcare training as such, and I was worried that they wouldn't have a bedside manner that would actually facilitate their ability to be good counsellors. But over the years I'm glad that I've been proved wrong in that, because I feel we've had some absolutely excellent candidates and graduates from those sorts of courses who do get a very good training in the counselling side of things.

The other thing I wanted to say was that there came a time, around 10 to 12 years ago, when a little bit of friction began to emerge between the medics and the counsellors about the counsellors potentially having their own caseload that would be independent and never seen by the consultant. It was actually Peter Farndon, when he was Chair of the Clinical Genetics Society, who said, 'We need to deal with this. We need to have some sort of working party to look at the demarcation of roles, essentially.' That was a group that I sat on, and actually it was pretty straightforward. We were able to work out quite clearly where the boundaries were, in fact, and where a genetic counsellor (GC) should not go, if you like, and which was territory for the medics. As part of that we did look at what was happening in other places in the UK, and in Europe, and the service, and the country that is perhaps closest to us is the Netherlands where their medics work in a similar way. They have rather a lot of counsellors, but in fact their counsellors do far less than our counsellors here and that's probably still the case, although I don't have an update on that.

Lucassen: I can give you an update on that. There are roughly about a quarter of the number of counsellors in the Netherlands compared with the UK, and they don't have an independent caseload like UK ones do. Medics are ultimately responsible for any patient referred, and this has led to a practice where medics have to pop their head around the door for any consultation that a GC holds because they are nominally responsible. So it's quite different here.

Hodgson: From the point of view of cancer genetics, the ability to deliver cancer genetics services did seem to correlate with how genetic counsellors were accepted into their role and given more responsibility, but again, it comes back to the issue of defining their role very clearly so that people don't feel threatened that somebody is going to impinge on their turf. Also, that they might make mistakes because they are dealing with something that they aren't trained to deal with.

Lucassen: Can I just ask: how do you feel then that genetic counsellors have impacted on the emergence of an ethical dimension in clinical genetics? I suppose that's what we're trying to get at really, isn't it?

Harper: Personally, I'd completely like to go along with what Nick Dennis was saying about how isolated one can feel without somebody to discuss things with and back one up, especially, I found, in something like Huntington's where stressful situations are abundant. We were also very fortunate in Cardiff to have a superb social work-trained person, Audrey Tyler, who was, well, she was just magnificent with the families, and she had an academic frame of mind as well.[107]

Just to go back right to the beginning, the situation in North America, which I didn't realise but it actually did develop rather differently in terms of genetic counsellors. When I was working in North America things hadn't really differentiated, things were at the stage that Angus mentioned; there was a nurse, or two of them, and other people, doing rather specific but different areas of work, often research-related. There weren't any defined genetic counsellors. Then the course at Sarah Lawrence got underway with the backing of clinical geneticists.[108] But at some point, and I only realised that very much later because I'd missed out on it, I'm very glad I had, a lot of tension had arisen in the North American situation, which started off on a funding basis as to who could be reimbursed for what, and then what happened, more or less, was the genetic counsellors who were starting to emerge got excluded because they couldn't be registered along with medical practitioners. Then the counterpart happened, which was when they formed their own society, clinical geneticists were excluded very specifically. So there was this polarization, and a lot of resentment, which I think has more or less diffused now. Fortunately, perhaps not absolutely entirely but very largely, we missed out on that in this country, because I think now everybody would realise that there are two complementary, and indeed equally valuable, roles and I'm just very glad we didn't have that period of bitterness.

Then, right at the very beginning, of course, we mustn't forget that the people doing genetic counselling were non-medical biologists, and actually one of the best books is by Sheldon Reed called *Counseling in Medical Genetics*, written in the 1950s.[109] He was a *Drosophila* geneticist, who, after the war, took a post in what was a sort of eugenic set-up, but he basically completely ignored this. He says so in his book. He was clearly, even though he had no training on the ethical and social aspects, he was clearly a very empathic person. There was a

---

[107] See, for example, Tyler and Harper (1983).

[108] Sarah Lawrence College, New York, offered the first postgraduate course in genetic counselling in the USA; see www.sarahlawrence.edu/genetic-counseling/ (accessed 6 October 2015).

[109] Reed (1955).

good study recently on him in a book by a person called Minna Stern, it's called *Telling Genes: The story of genetic counseling in America*.[110]

Modell: I would like to raise the question of who does genetic counselling because, as genetic knowledge expands, more and more clinicians really are in a situation where they need to provide genetic information. I'm aware of a couple of studies that were focusing on the transmission of genetic information in British Pakistani families. One was in Blackburn and one was in Bradford, where the researcher visited families with known genetic diagnoses for which the DNA basis was known and carrier testing was feasible, and asked where they got their information from.[111] It turned out that a minority of these families, both in Blackburn and in Bradford, had actually seen a genetic counsellor. Where they had their information was from the paediatrician. Now this was a particular ethnic group. Why hadn't they seen a genetic counsellor? Had they been offered? Had they not gone? Had they forgotten? I don't know. As far as they were concerned, the person who provided them with genetic information was the paediatrician. And that, when you consider what's required, the timing that's required and so forth, the skills, you do feel that community paediatricians really need a new group of community genetic counsellors who can work with them, the community and clinical geneticists. So when we're talking about genetic counselling, you were talking about the development of the profession, but I think there's a huge unmet need.

Lucassen: I think it's a very good question, as well as 'what is it that makes someone a genetic counsellor?' We've described genetic counselling in the context of clinical genetics, haven't we, and you're quite right that the demand for genetic counselling also happens outside the profession.

Modell: But it's getting, it should be getting more and more recognized that every clinician needs to have basic genetic counselling skills.[112]

Lucassen: It's also still not uncommon for a patient to ring up and say, 'I'll have the genetics bit but not the counselling, thank you.' So the term counselling also has very different interpretations, doesn't it?

Hallowell: I just wanted to make an observation, which is three of you at least have said that you find it very lonely doing genetics consultations, and I wondered

---

[110] Minna Stern (2012).

[111] Ahmed, Green, and Hewison (2002).

[112] See, for example, Emery and Hayflick (2001).

whether that is because there is something specifically difficult ethically about genetics, coming back to the question about genetic exceptionalism before. Why do you all feel so lonely when you're doing the consultations?

Dennis: I think it comes down to having not had any training in how to talk to people really. You're giving people bad news, and it took me many years to have some inkling of the best way of doing that. I was self-taught really, and I think by the time I retired I wasn't too bad at it but the first half of my career I was pretty bad at it. I would arrive at a point in the consultation where I'd think, 'What the hell am I supposed to say next?' I just wish I'd had somebody else there who could have looked over my shoulder and made some suggestions.

Hallowell: Lots of other specialties give bad news, so is there something special about genetics? Is it because there are especially difficult ethical issues?

Dennis: I think it's just because I qualified in 1968, and most of the people around now qualified later than that.

Lucassen: But there would be oncologists who qualified in 1968; would they also describe this loneliness? I think Nina's point is well made. What is it about that situation that I recognize, but I can't quite put my finger on what it is?

Dennis: The nature of what you're saying in genetics is different from what might happen to you because of this illness. I mean you've come because you've got an illness, so we all acknowledge that you've got an illness. So that's the difference from when 'well' people are coming to you and perhaps getting information about what might happen in the future, or what might happen to their children. It's a different order of knowledge really. I think that does make it harder. You're presenting them with a situation that perhaps they hadn't thought about before. They don't know how to react to it and it's quite difficult sometimes to think how you ought to, well to think how you can, help them.

Hallowell: Martin, we'll do it together: Is it the ethical thing or is it a family thing?

Richards: Carry on.

Hallowell: I was just trying to get at if it is any different from any other area of medicine, and is that because it's an ethical difference or is it just a technical difference maybe, because it's certainly not a technical difference. Do you think there are special ethical issues, really, within genetics?

Pembrey: I'm not directly going to answer that question, but the first thing to say is that in all my practice – I stopped clinical practice at the end of 1998 – we never in discussion referred to Beauchamp and Childress' four principles, we never used the word ethics particularly. What we were worried about was delivering our professional duty of care, and in other parts of medicine the professional duty of care focused largely on the individual was much better defined than with clinical genetics initially. We were feeling our way, particularly in the 1970s and 1980s. And just to talk about giving bad news, in 1966 when I was a house officer, I told a patient who was having a biopsy for what might be Hodgkin disease, I think, I promised that I would tell her the result when she came back from the operation. My consultant was horrified. He said, 'You can't just take it upon yourself to do that, but if you've promised…' I said, 'Well, I'll deal with it', and he said, 'Well, I've seen the hope drain out of their faces when you give them that news'. That would seem amazing now, but in developing the clinical practice in clinical genetics we basically just hung on: 'What is our professional duty of care? How are we going to help not just this couple who are coming, or this person, but the family in general?'

Lucassen: The lived ethics that Nina was talking about.

Pembrey: Maybe that's what the academics call it but we didn't call it that.

Lucassen: No.

Pembrey: We were very definitely engaged in protecting what we had built up for these families at high risk. In 1987, for example, when there was a parliamentary debate on MP David Alton's Bill on whether the legal age of termination of pregnancy should come down to 18 weeks, and what that would mean for these families; Bernadette (Modell) and I were travelling back from a meeting in Cardiff – that would have been in 1985, 1987, perhaps 1986, something like that – and anyway, it was a meeting in Cardiff. Leading up to this parliamentary debate, the pro-life lobby, the anti-abortionists, had a phrase, a catchphrase, 'search and destroy'; that is, all that prenatal diagnosis and counselling is search and destroy. We had to work out a phrase, and we decided what it was by the end of the train journey: what did we deliver? We 'restored reproductive confidence'. That phrase came out of the work that Bernadette had done on thalassaemia, showing in the Greek Cypriots of northern London that if they didn't know anything at all, if they just had children, and lots of them, with Mendelian ratios, some of them would have the disease and die.[113] If they were given genetic counselling but nothing else, they were too frightened to have children. They stopped having children, 73 per cent,[114] and so what we were doing by introducing prenatal diagnosis was restoring reproductive confidence; so all that went on.

We were dealing with issues that were there at the time in a practical way, worrying about professional duty of care. With all due respect, I went on a Raanan Gillon course on ethics and philosophy for medics, and so on, and they had a variety of medics.[115] When I arrived there they said, 'Oh thank God we've got a medical geneticist, a clinical geneticist, with some real examples of ethical issues'. They were desperate for practical examples, you know. So they were interested, and they called it 'ethical issues' but we didn't use the phrase. Maybe I'm out on a limb on this one.

Lucassen: But does that matter that you didn't call it ethical issues? You're describing ethical issues, does it matter that you didn't call them that?

Pembrey: No, no, it doesn't matter at all but we developed things on the hoof a bit, to be honest, with the professional duty of care being the thing, principally, that we talked about.

Modell: And listening to your patients. I was thinking that the way the ethical approaches to new technical developments happened was when you informed your patients about them, you listened to their reactions and those were so deeply educational that they then led on, I think, in most practising clinicians, to a certain type of ethical practice, very non-directive basically. Then it was an empirical collection of the attitudes of those clinicians that led on to the defining of fundamental principles of genetics.[116] But the bottom line is the patient, the consultation, the patient interaction, and being influenced, being open to what the patient said.

I'd like to just add one final thing about loneliness. I did understand when you were talking, I realised what you were talking about because there was a breakthrough moment for me at the time that I was involved with the patients

---

[113] Modell and Berdoukas (1984); see Chapter 4, 'Thalassaemia in Britain', pages 76–85. See also Professor Bernadette Modell's account of this work at a previous Witness Seminar, Jones and Tansey (eds) (2014).

[114] Modell, Ward, and Fairweather (1980).

[115] See note 46.

[116] See Professor Bernadette Modell's comments on page 21.

in founding the UK Thalassaemia Society Support Association and we had some meetings that were quite emotional.[117] In one of those meetings I suddenly realised and said, 'I understand now. You don't, you're not expecting me to solve your problem, you just want me to be there for you?' and they said, 'Yes'. That was a great relief, and I certainly didn't feel lonely after that. Does that address part of what you were saying about loneliness?

Dennis: Yes, that is part of it, it's the feeling that you want to make everything right, and you can't. But I think the main source of difficulty for me was not that there were ethical problems involved in some of these consultations, because I think that's not the most difficult part of it. I think it's the fact that they didn't know what they were coming for quite often. They were not ill and you were telling them something that was in a mental space that they perhaps didn't even know existed, and so I think that added to their difficulties and their distress. As a doctor, I did want to make that right, and I couldn't see how to make that right. I didn't have time to get on the phone the next day, or go and visit them at home or anything like that, but I had this idea that there might be somebody who might be able to do that.

Modell: Yes, and that comes back to what Peter was saying about the origin of working in medical genetics when there wasn't much you could do, and you wondered why were people so welcoming and so pleased to see you. Because you were there for them.

Parker: I was just going to respond to Marcus' point from the other perspective, from the other side, from the philosophy side in a way, about the involvement of humanities and social sciences. In the second half of the 1980s I was doing my PhD and I was living in London, and I was running, with Don Hill, a seminar series for the Society for Applied Philosophy, so workshops essentially on a Saturday morning, and anyone who wanted to could come.[118] I remember there were several of those over the years that had something to do with genetics in them, and I remember sessions led by various people. But if I'm really honest, they were often wonderful speakers and I loved their work, but the issues didn't grab me; I didn't find them that interesting. But something happened later on, a couple of things happened, which changed it for me. In 1998 I had a fellowship in Melbourne, and I did a bit of work with Bob Williamson at the Murdoch

---

[117] For the UK Thalassaemia Society, see http://ukts.org/about-the-ukts.html (accessed 30 June 2015).

[118] For Don Hill, see, for example, Almond and Hill (eds) (1991).

Institute, so I spent some time there.[119] I was asked to facilitate a meeting of genetic counsellors and geneticists, and there were some interesting cases. Then the second thing was, when I moved to Oxford in 1999, I met Anneke and this was on a one-week medical ethics workshop where clinicians brought cases or issues, things they wanted to turn into papers. I remember Anneke brought a case about non-paternity. Both of those struck a real chord for me because the thing I'd found uninteresting about analytic philosophy and its approach to ethics was that it didn't touch on the things that I thought were really important in ethics, so the lived experience of people struggling with how to live and how to make difficult moral decisions is what really grabbed me, and there was none of that in these fairly abstract philosophical talks. Those two experiences, being in a genetics clinic in Melbourne and working with you, Anneke, on that paper around this difficult decision, really made ethics much richer for me and made it much more interesting and engaging. At that point for me, personally, obviously other people's stories will be different, I was always socially minded as a philosopher, both politically and theoretically, but those two things came together, and the prism through which I saw philosophy ignited in a really interesting way.

Lucassen: I think that's really interesting because I think what I've been hearing is that ethics is conceived of in different ways and some people talk about formal ethics or lived ethics. For me, it's just an echo of what you're saying, is that I didn't really know how to apply formal ethics or frameworks, or whatever, in my clinical practice. I couched ethics as good professional practice. But somehow meeting Mike and talking through things in meetings with him, in a lived way, made ethics in practice seem very different.

Parker: Yes, I think there's a lot in common. I think it's very important; I mean the term 'formal' is quite problematic there I think because it suggests, without explicitly saying it, that there are other ways you can think carefully, systematically, and thoroughly about moral problems and engage with them, not just thinking obviously. There's a whole range of things going on, so I wouldn't want to suggest what we're talking about is 'informal'; I think there's a discipline to it, and it's real and engaged and rich. It's not formal in a sense that I don't think even Beauchamp and Childress really saw it as either, but it's the way they tend to be interpreted.[120] So to say ethics is not formal doesn't mean it's not serious or engaged, I don't think.

---

[119] Professor Bob Williamson was Director of the Murdoch Institute and Professor of Medical Genetics at the University of Melbourne (1995–2004).

[120] Beauchamp and Childress (1979).

Lucassen: No, that's a very good point.

Mumford: I want to add something, also in response to Marcus, and to introduce another dimension. In 1990 I was asked to be the secretary of a new ethics committee that would be examining the issues arising in the 'Children of the Nineties', or ALSPAC (Avon Longitudinal Study of Parents and Children).[121] I don't know if you're all familiar with that? It was an epidemiological study of 14,000 mothers and 14,000 babies. To me, the initiation of a study of this scale raised a challenge to the question: 'Can you just do it on a case-by-case basis?' Clearly, when you're asked to set up ethical principles for a study like this you have to do something overarching before you start looking at individual cases, and that's in fact what our ethics committee did.

What I think was unique at the time, although now it's more common, was that our committee was asked to formulate the principles that would guide the study as a whole – then respond to them as things came along. In setting up the principles, we had to have some sort of common policy. On our committee of ten there was one philosopher, a theologian, and a couple of lawyers. Most of the others were doctors and scientists, so not a lot of people with a formal background in medical ethics or in moral philosophy. Nevertheless, we had to sit down and draw up principles, and use them in application in the study. It was at least in part a genetics-based study but a very different sort of thing from the clinical practice, or even from the sorts of clinical trials in which people here might have been involved, because of its size and scope, the nature of the questions we were asking and the samples that we were looking at.

Lucassen: Can you say a little bit about the sorts of principles that you came up with?

Mumford: On the first meeting, perhaps the first two meetings, we decided the essential issues to be addressed. These were confidentiality, consent, and the reporting of results to participants. Confidentiality, I think, was more or less taken as given at that point, at least as far as the genetic side of things was concerned.[122] The biological samples we were looking at were maternal and cord blood, and

---

[121] ALSPAC began in 1991 at the University of Bristol and is an ongoing study; www.bristol. ac.uk/alspac/about/ (accessed 30 June 2015). The Study was also the subject of a Witness Seminar; Overy, Reynolds, and Tansey (eds) (2012).

[122] Mrs Elizabeth Mumford wrote, 'There were issues that arose in terms of other parts of the study: questionnaires and face-to-face meetings, but as far as the biological samples were concerned, it seemed pretty clear.' Note on draft transcript, 12 January 2015.

maternal urine. The genetic testing was done on the maternal and cord blood. It was clear to us that we owed a duty of confidentiality to the subjects.

Consent was a more difficult issue because of the prevailing guidance at the time. In fact, when we were drawing up our principles, most of what would become prevailing guidance hadn't yet been published. But the guidelines that came soon afterwards from the Royal College of Physicians, for example, said that if you were doing, as we did, and simply taking for research use an additional small amount from samples that were already being collected for clinical purposes, there was no need to seek either consent from the patient, or even ethics committee approval.[123] Yet, at the same time, some of the advice that was coming out about genetics was that, if you were doing genetic testing, it was necessary to seek individual consent for every single test. You must seek individual consent, but possibly not if the sample was anonymized.[124] In time we had to grapple with the apparent contradiction between these two points of view, but, of course, that was in the future. At this point we had no professional guidance. On the committee's conclusions, it's interesting, the chair was an academic lawyer, a very eminent academic lawyer, Professor Michael Furmston, and the very junior secretary was another academic lawyer – that was me.[125] Never underestimate the power of the minutes' secretary! [Laughs]. What you write up really matters. We took the view that, although we didn't think we had a legal duty to seek consent, we felt there was an ethical obligation to do so, and so we did. This is the wording that we used in the letter that was sent to participants: 'We will be keeping some of the blood and urine that is routinely taken during pregnancy. We will keep the placenta once the baby is born. With the blood we will be able to find out whether you have an allergy and whether your genes affect the health of your baby.'[126] So we got the word 'genes' in, but it was surrounded by a lot of other material because we didn't want to terrify people. We thought people were afraid of genetics, and if they saw the word they'd think, 'Oh, they're going to clone us!' [laughter] or 'They're going to find out that the paternity of the baby is something other than what we'd expected'. But we felt we had to say something about genetics, and there was a formal, written consent for the use of biological samples.

---

[123] Royal College of Physicians (1994).

[124] See, for example, Nuffield Council on Bioethics (1993).

[125] Michael Furmston was Professor of Law at the University of Bristol from 1978 to 1998, now Emeritus.

[126] Quoted from the initial information that was sent to participants in 1990, entitled 'Joining in Children of the Nineties: Questions and Answers'. Email from Mrs Elizabeth Mumford to Ms Emma Jones, 11 November 2015.

Although those samples were collected during pregnancy, because that's when it was routinely done, we felt that this was not an appropriate time to be asking people for consent to the study, because they were being asked for consent about things like alpha-fetoprotein, which they really needed to think about because it might have a bearing on their own lives. So the consent was actually done some considerable time after the samples had been collected. We told the participants that we would not be using the samples without their written consent. Again, the words we chose mattered very much afterwards: '… but we won't do any of these tests unless you give us your permission', which meant that we had all those samples stored and ready to use, but nothing could be done until much later when these people had been tracked down, and consent either given or refused. The other thing that I just found myself saying, which is worthy of noting, is that I'm saying 'we' all the way through my remarks. Although I was not doing the study myself, this gives you an idea of the way in which we all got very caught up in it, and we really felt we were part of the whole study.

Clarke: I just want to say a couple of quick things: one was to go back to the loneliness thing. In addition to all the reasons why we feel the weight of what's happening with the families, certainly for a phase of a good decade or more, a lot of people were very lonely in clinical genetics because they were running units on their own. They didn't have ward rounds, and they didn't have people coming in and out of the clinic room. You were driving off to a unit 100 miles away to do a clinic and then driving back again on your own, and you were on your own out there too, so there are lots of reasons for feeling lonely, I think. Coming back to the train of thought about feeling the weight of things for the families, and I think something that was very influential for me and I think for a whole generation of people coming through clinical genetics was the experience of getting involved in linkage studies during the 1980s. These meant going out into people's homes, maybe hundreds of miles away, seeing half a dozen people in a day. It meant going off to a village and rounding up members of the family affected, unaffected parents, at risk, all these people, and spending a day with them and then going off and doing the same the next day, and the same the day after that. So you had a real education in family dynamics and the impact of major genetic disease on those who are both affected, at risk, parents of, etc., etc. I think that affected the way a lot of us still around look at things. Things happen differently now [laughter] because of the technology, but that was what happened at that time.

Fryer: Well, I'd echo exactly what Angus has just said about linkage studies and traipsing round the country. I think it's extraordinarily valuable. Going back to the loneliness issue and Nick feeling untrained. I can't say when I went to Liverpool that I was untrained because I'd worked with Peter for three years, but I recognize the feeling of impotence when you're seeing a patient and you may be giving them bad news. Unlike when I was a paediatrician, and you'd possibly say, 'Well, I'll see you again in clinic in six weeks' time', and you were following them up, you just did not have the resources to do that as a clinical geneticist when you're looking after a population of two and a bit million. One of the tenets of our specialty is providing information and support. I think it's the support bit that is very difficult if you're on your own without any other professionals, without any other doctors often, but certainly without any other professionals. The genetic counsellor had many different roles, but one of them, certainly when I went to Liverpool, as a virtually single-handed consultant, was in providing support; that kind of follow-up to families was a really critical part of the genetic counsellor role.

Lucassen: I think that's really interesting because there's information and there's support, and if the genetic counsellors were doing mainly the support then perhaps it's not surprising they've been more wrapped up in the ethical issues if indeed that's true.

Hallowell: Can I just say, to make you all feel better, I have, as a social scientist, interviewed many hundreds of people who have been to genetic clinics and they love you [laughter]. They all feel completely supported and it is absolutely always a very positive experience for people, and I do worry on that basis. I mean, a lot of the patients that I've seen are going mainly through cancer genetics clinics, have very long consultations compared to your average consultation in the hospital and I think that will not last forever and a day, and these issues are very complicated. I'd like to say I think you've all been doing ethics ever since the very beginning. Well, how I see ethics. I do wonder how this will be carried forward into the future because there just aren't the resources to provide this kind of service. So I just thought I'd make you all feel better.

Clarke: We don't give you research access to our failures. [Laughter]

Lucassen: That was a conversation stopper, Angus.

Bale: I just wanted to pick up on something that Nina said. I totally agree with everything I've heard, and I thoroughly enjoyed working with all the clinical geneticists, but I just have a sense of a 'but' that I wanted to bring into the conversation. Partly through my experience when we established the Human Genetics Commission (HGC), which is on the agenda, and I'm conscious that you and several other members are still here, so I'm feeling a bit like Elizabeth (Mumford) in being the secretary to the HGC.[127] There was a conscious decision taken early on to get a mixed membership, very much lay-driven. I think what was interesting about that, I looked at it from behind the scenes, it was actually quite quickly that two groups emerged and we saw it particularly when we tried to get a panel of patients together. One panel whom you might call the rare diseases community, who came with a very much worshipful feeling about genetics generally, and one from a disability rights background, who came at it with quite a different perspective. I think it was always very interesting to have this debate, and I'm sure Martin (Richards) will recall where there was quite a lot of tension about whether we should be dealing with issues around reproductive choice and all those sorts of issues. So I think there is a quite interesting angle on this one which has slightly gone away, whether it's just gone away from my radar but it doesn't seem there's quite such a polarized debate about it.

Lucassen: You mean the Human Genetics Commission has gone away?

Bale: Well, that's gone away, I know, but I mean without HGC, I don't get views from members such as Bill Albert and other opponents of genetics constantly lobbying me anymore.[128]

Lucassen: Can you say a bit more then about the Human Genetics Commission?

Harper: And particularly say how did it come about given the reluctance of Department of Health (DoH), initially, to have anything to do with something like that?

Bale: Well, we've talked about this and I've been fascinated by Peter's attempt to try and get the history of medical genetics together.[129] We haven't really had a proper look in the DoH files. I suspect a lot of what happened before, probably in the mid-1990s, was all around the Royal Colleges and maybe around a

---

[127] The Human Genetics Commission ran from 1999 to 2012, and was the advisory board for the UK government on human genetics-related policy issues; see http://webarchive.nationalarchives.gov. uk/20121102204634/ and http://hgc.gov.uk/client/content.asp?contentid=5 (accessed 30 June 2015).

[128] Dr Bill Albert was a member of the Human Genetics Commission from 1999 to 2005; http://webarchive. nationalarchives.gov.uk/20060810161251/http://hgc.gov.uk/client/content.asp?contentid=661 (accessed 7 October 2015). See also Albert (2007).

[129] See note 7.

little bit of work that might have, I think Peter thought it might have, been paediatrics. What really sparked the interest in genetics in the department and I was part of that, which is probably why I have a post, was two things. One was the hype about gene therapy, which we shouldn't forget predated some of the work around the Human Genome Project.[130] From the early 1990s there was a Gene Therapy Advisory Committee, which predated all the other genetics commissions, and it was interesting because, of course, it brought in a research ethics dimension to some of the debates around genetics and what we might be able to do.[131] It was a very interesting committee when I first got involved in it, not as a DoH employee but as someone from outside in the Health and Safety Executive, in that it almost felt like it was making decisions about treating individual patients, something an ethics committee wouldn't do now. Subsequent to that, and I think Peter and I were reminiscing the other day, there was a very important report, which I must go back and read, by the House of Commons' science and technology Select Committee. It reported in the mid-1990s – I think the report was just called *Human Genetics* – and I've always been very impressed with that inquiry. I always keep meaning to go back and look at that report, look at who was behind it, who the experts were.[132] What was intriguing about it was that the recommendations were made, a government response was published, and one or two of the responses annoyed the committee, or one or two members of the committee. They did something that is quite unusual and one of the things that you never want to happen, which is they actually brought out a supplementary report in effect criticizing the government response.[133] That then led to the establishment of

---

[130] See, for example, French Anderson (1989).

[131] The Gene Therapy Advisory Committee was first convened in 1989, chaired by Sir Cecil Clothier QC; see Taylor and Lloyd (1995).

[132] House of Commons Select Committee on Science and Technology (1996). The members of the committee were Mr Spencer Batiste, Dr Jeremy Bray, Mrs Anne Campbell, Cheryl Gillian, Dr Lynne Jones, Mr Andrew Miller, Mr William Powell, Sir Giles Shaw, Sir Trevor Skeet, Sir Gerard Vaughan, Dr Alan Williams. Specialist advisers were Professor David Porteous, MRC Human Genetics Unit in Edinburgh, and Dr Bryan Sykes, Institute of Molecular Medicine in Oxford; see pages ii, xviii. For details of the 42 witnesses who testified to the committee, see pages v–vii.

[133] For the Government's first response, published in January 1996, see Department of Trade and Industry (1996a). For the second response, published in June 1996, see Department of Trade and Industry (1996b). A copy of the latter, which is not widely available, is held at the Parliamentary Archives, www.portcullis.parliament.uk/calmview/ (accessed 11 August 2015). For further commentary on these responses, see Select Committee on Science and Technology (2001).

two committees, if I recall correctly. One was called the Advisory Committee on Genetic Testing – with the acronym ACGT, the four bases of DNA. The other one was set up slightly later, in what was then the Department of Trade and Industry, called the Human Genetics Advisory Commission, which Baroness Onora O'Neill chaired.[134] John Polkinghorne chaired ACGT, and I'm sure there are probably people here who were on ACGT.[135] I was brought in as secretary and I attended the last meeting to wrap it all up. Then it was replaced by the HGC, which started, it's becoming a very long-winded answer to Peter's question but it started, because the new Labour administration had a desire to tackle a number of areas where they thought it had gone very badly on things like GM food, and to really engage with the public.

One was on agriculture and biotechnology, the AEBC (Agriculture and Environment Biotechnology Commission), and that didn't really have a very comfortable time. HGC, I think, had a slightly easier time because it was tackling issues that people could see the benefits to them, personally, of genetics. I think what was novel about the Commission as well was that it was an intent to be independent and to work very much in the public domain, and actually to go out and try and engage with the public, which, many of my colleagues used to say, 'That's a very brave experiment. Good luck on that one'. We had some mixed signals there, so I personally found it a very rewarding part of my career and I look back on it with very fond memories. So that's the background to it, broadly.

Lucassen: It also very deliberately went out and tackled some of the ethical issues that came up in genetics, didn't it?

Bale: Yes, I think in a way ACGT was an interesting committee when I look back on their reports because they tried to put guidelines in place and codes of practice, and voluntary codes and so on, and they also went out and did site visits, which sounded a bit like CQC (Care Quality Commission) inspections from the people who reported back. You actually would go to a genetic service, and I don't know

---

[134] For a debate in the House of Commons in which the foundation of the Human Genetics Advisory Commission and the Advisory Committee on Genetic Testing are discussed, see HC Deb 19 July 1996, vol. 281 cc1405–78; http://hansard.millbanksystems.com/commons/1996/jul/19/science-policy-and-human-genetics (accessed 1 July 2015). The Commission ran from 1996 to 1999, when its functions were included in the Human Genetics Commission; see http://webarchive.nationalarchives.gov.uk/+/www.dh.gov.uk/ab/Archive/HGAC/index.htm (accessed 1 July 2015). The philosopher Baroness (Onora) O'Neill of Bengarve was Principal of Newnham College, Cambridge (1992–2006), and Chair of the Nuffield Foundation (1998–2010).

[135] See Professor Marcus Pembrey's comments on pages 69–70.

if anyone experienced them, and talk to people and almost peer review the service they were providing. But it tackled some of the ethical issues, and the Human Genetics Commission looked at clinical genetics through a very different lens, I think, with a much more lay perspective and tried to engage the public. Some of the early work on genetic privacy, which I forget what it ended up being called, *Whose Hands on your Genes?* [136] No, that was the consultation, and then it was *Inside Information*, I think, was the report. [137] They tried to lay down, and I still look back occasionally for guidance on some of the points because they did lay down a very useful, broader set of ethical principles. I remember when Mike (Parker) was chairing some work on the 100,000 Genomes Project going back and looking at the report and trying to see if those frameworks would work for where we were then because it was a clinical genetics approach, and now we're looking at genomics. [138]

Richards: Perhaps I could just slightly broaden the context because there are some things we haven't talked about at all. This is from my own personal perspective, but I became involved in working in genetic clinics in the mid/late 1980s. Two things I'd point out about that era: one was that I remember being very surprised, I mean having spent most of my life trying to get money out of funders to do research, that the MRC (Medical Research Council) practically came and gave us money. They were very interested in funding work around genetics because they clearly saw it as being developed and, as it were, producing new issues, and quite specifically funding social science work in that area. The other thing that's not been mentioned is the Nuffield Council because their very first report, if I get it right, was on genetic testing. [139]

Lucassen: 1993.

Richards: Some of the next few were also on genetic issues and it seems to me they were important, and someone who travels through some of those debates in an important way is Onora O'Neill. [140] I do think we have to think

---

[136] Human Genetics Commission (2000).

[137] Human Genetics Commission (2002).

[138] The '100,000 Genomes Project' intends to sequence the genomes of patients of the National Health Service with rare diseases, members of their families, and those with more common cancers by 2017; www.genomicsengland.co.uk/about-genomics-england/ (accessed 1 July 2015).

[139] Nuffield Council on Bioethics (1993). The Nuffield Council on Bioethics was established in 1991, and has addressed topics such as genetic screening, donor conception, and genetics and behaviour; see http://nuffieldbioethics.org/about/ (accessed 7 October 2015).

[140] See, for example, O'Neill (2002), Manson and O'Neill (2007), and also note 133.

about what research was getting funded. The other thing, a little bit after that, of course Wellcome set up its panel specifically on, what was it called, ethical issues, yes, but much of that was actually supporting work in relation to genetics in one sense or another.[141] So there was a whole generation of PhD students getting funded on those programmes themselves, the whole increase of social sciences people, who had an interest in ethical issues. Lots of it actually focused directly on genetics. I mean, that was the central issue in all of that, I think it's fair to say.

Harper: I think HGC was certainly successful in getting not just the problems of genetics but some of the ethical issues out to communities that weren't familiar with them. I recall in particular, I think you were on it, Mark (Bale), a visit to the National Forensics Police Centre in Birmingham where we were duly shown around, and I remember somebody on the visiting group, when they had been describing all sorts of wild projects on potential criminality and things, somebody asked, 'Now, what about your ethics committee?' And there was a sort of dazed silence at the idea of an 'ethics committee?' It wasn't that they just didn't have one, it never even crossed their minds that there might be any need for ethical review. Then, when we did manage to get down to the question of the ethical oversight, they said, 'Yes, there are three people who are responsible for overseeing it', and I can't remember the official titles. It turned out all these three people, 'they' were actually just one person who was in charge of it all. Now I don't know if that made any difference, because the whole thing was privatized shortly afterwards but they certainly had their eyes opened and that was due to HGC. I hope that's somewhere in the minutes, Mark.

Bale: Yes. I think it's a very interesting example, and I think it also showed in some ways the advantages of having Baroness Helena Kennedy chairing the committee because she saw DNA in a different way.[142] I think you are absolutely right, there was one guy who was custodian and the kind of gamekeeper and poacher, but they did in the end set up a national oversight committee and the HGC was represented on it. I've lost track of where they got to but that did really seem to open our eyes, I do recall, in that meeting.

Pembrey: If I could go back a little bit to the Advisory Committee on Genetic Testing, because, with Peter, I think we were both on that committee, I was from 1996 to 1999, or something like that. One of the things I was landed with was

---

[141] The Wellcome Trust's Biomedical Ethics Research Programme started in 1997.

[142] Baroness Kennedy QC chaired the Human Genetics Commission from 2000 to 2007.

chairing a subgroup on over-the-counter testing and we actually published in September 1997.[143] There were two things that came out of that; the first thing is that when people say that the ethical and other types of practical considerations were 'follow-on' to the science that races ahead and so forth. This was the exact opposite. It turned out that we'd got all these principles and everything else, we had a press conference, we had a leader in *The Times*.[144] I was on the *Today* programme and everything else.[145] It turned out there was only a single company offering cystic fibrosis testing, and they folded in about six months, and it was long before '23andMe'.[146] Of course, the whole thing has been revisited with advances in genome sequencing.[147] That's the first thing.

The other thing I remember, and it brought into sharp focus this thing that we've been discussing, about whether there's anything special about genetics compared with other issues, was that Matthew Parris, *The Times* journalist, was on that committee, and when we were saying, 'Well, we've got to be so careful when this information is handed over because it doesn't have implications for just the person…', he said to me, 'What's all the fuss about? People can go and buy a pregnancy test over the counter! You know, what could be more important than whether they are pregnant or not?' Our answer, of course, was, 'Yes, people know what pregnancy is, and it will become evident very shortly if the test is correct or not,' and so on. We then had to try and argue the point that genetic information, carrier testing, or presymptomatic testing in some situations, was different.

---

[143] Sub-committee on 'Over the Counter Genetic Testing'; see Advisory Committee on Genetic Testing (1997).

[144] Hawkes (1997).

[145] Daily current affairs programme on BBC Radio 4.

[146] '23andMe' is a private company selling DNA self-testing kits that use saliva samples; see www.23andme.com/en-gb/ (accessed 1 July 2015).

[147] Professor Marcus Pembrey wrote: 'Genetic testing has been revisited by the inclusion of the NHS as well as commercial genetic testing in the discussion. I was mindful of the UK White Paper on genetics in the NHS, Department of Health (2003). This triggered several reports, including *Making Babies: Reproductive decisions and genetic technologies*, Human Genetics Commission (HGC) 2006; and the HGC and UK National Screening Committee joint report on *Profiling the Newborn – a prospective gene technology?*, Joint Working Group of the Human Genetics Commission and the UK National Screening Committee (2005). Then there were a huge number of publications on the ethical aspects (including what to do about "incidental findings") as genome analysis techniques advanced. Google funded "23andMe" in May 2007.' Note on draft transcript, 27 August 2015.

The other thing I want to raise is that I became the consultant adviser in genetics to the Chief Medical Officer in 1989,[148] and so there was a question of trying to establish some general principles, ethical and otherwise, for *Population Needs and Genetic Services*; Ian Lister Cheese and I worked on this document with the help of a lot of others.[149] We produced it, and in, let me get the piece of paper itself, it was sent out in 1993, a letter from the Chief Medical Officer and also from the Chief Nursing Officer, and it was distributed very, very widely. The booklet itself has no identification on it as to whether it was from the Department of Health or anything. The printing is HMSO (Her Majesty's Stationery Office) but beyond that there is nothing as to where it's come from, so as soon as it separates from the covering letter, and I said, 'Surely this is a bit odd?' and I think Ian said, 'No, it was just an error.' But part of me thinks that, like the advice when I first came to the consultant adviser post, the question of prenatal diagnosis and selective abortion was too hot to handle when I arrived. This was very definitely something they wanted to not have to make a decision one way or another about. I think the one thing that it brings back to me is that the selective abortion side of things is something that really drove the engagement of all the ethical issues and so on, more than anything else, during the 1980s and 1990s.[150]

Farsides: I agree with you. I can give an example of another interesting silence on that issue. I had a student who was a very senior midwife and she was interested in termination of pregnancy on the basis of fetal anomaly, in fact identified during scanning rather than during any other form of testing. But she did a survey of all the leaflets that were available for women in antenatal clinics and couldn't find one that mentioned termination of pregnancy. So there was this absolute separation between the idea that we can now give you this information and any suggestion that this is where it might lead in terms of the choice that you would make on the basis of that.

Hallowell: Sort of related but unfortunately it doesn't follow on very well: I was just sitting here thinking about what Martin said about the funding opportunities. It's clear, as far as social science is concerned, and its contribution to this area, funding opportunities like, for example, the Wellcome Trust programme, which involved taking young social scientists, philosophers, lawyers away to Hinxton, feeding them with a lot of very nice food and giving them workshops for two

---

[148] Professor Sir Donald Acheson was Chief Medical Officer of England from 1984 to 1991.

[149] Department of Health (1993).

[150] Ciba Foundation (1985).

or three days. These happened, I think the beginning was in the mid-1990s, I certainly was involved as a member of faculty in 1999 and 2000. And actually nurturing a whole group of these young academics must have had a very big impact on the sort of research that has been carried out and also the way in which the research is fed back into the clinic. Therefore, there is actually a translational sort of circle going on here, so lots of people were doing research on lived, ethical issues and translating it. In my research I always tended to go back to the clinicians; that's why I know a couple of you in the audience and actually talk about my findings and how maybe information could be delivered in a different way or how families could be approached and things like that. So I suspect what I'm actually trying to say, the research is obviously affecting clinical practice but fundamentally the funding opportunities for research is going to be affecting those sorts of things.

Parker: I'll say something about the Genethics Club, I guess.[151] One of the things I just wanted to pick up on is these issues, three of them in particular are enduring. They can't be resolved for all time because these are real people, real families, and real decisions, so they're always going to be around in one form or another. Reproductive decision-making continues to be one of the main issues in about a third of the cases; genetic testing of children is an issue in about a third of the cases, and about half the cases are to do with sharing information in families, or not sharing information in families. It's got to be more than 100 per cent because sometimes there's more than one issue. Reproductive issues are particularly difficult ones for people to talk about and find ways of articulating; for example, reservations that a health professional or genetic professional might have about prenatal testing for what they consider to be a minor condition. You know, how do you talk about that and think through that issue when there are all sorts of taboos and anxieties about even talking about those things?

One of things that struck me about my own experience in thinking about ethics in genetics is just how rapidly, I can date it almost quite precisely, how rapidly it kind of took over my life and has become an issue across the field in a sort of more visible way. I arrived in Oxford and met Anneke and we set up a monthly discussion group in the regional genetics service in Oxford where difficult cases were discussed. No funding or anything like that, and that's happened every month since 1999, so that's a long time; and that's essentially become an important thing. Very quickly after that other people in other units heard

---

[151] See pages 4–5, see also Parker (2012).

about that and asked me and you to go and do those, so I ended up doing one, I think we went to Cardiff, we went to Newcastle and had similar types of discussions. Then, Anneke, Angus, myself, and Tara Clancy in Manchester got some money from the Wellcome Trust for a one-day meeting, essentially to bring together people working in genetics across the UK to just talk about the kinds of ethical issues that were arising in their work because it was clear that people were struggling with these things individually but also that there were differences; the different units were struggling with things in different kinds of ways. And several of the people here now were at that first meeting. I remember it really well because I decided we'd go around the room and get people to introduce themselves and say why it is they thought ethics was important in genetics, and we had a whole agenda for the day, but we spent the whole day going around the table. People talked about cases, and we recorded the day. I've still got the transcript of that, and it was incredible because it was so rich and so interesting.[152] At the end, someone said, 'There will be a lot of value in this, running this again.'

Lucassen: Karen Temple, I seem to remember, said that.[153]

Parker: It may have been Karen. And so we set up, and that was 2000, and since 2000 we've had one of those, three of those, a year and they have moved around the country, no funding again. They have been attended by more than 800 people in total, about 400 completely separate individuals – some people come on a regular basis. We've had an attendance for between 30 to 40 people each time, and about 450 cases, I think, presented, formally presented, and, of course, that leads to other discussions. I mentioned the themes that come out most. There have been times, I remember, when several of us have said, 'Do we really want to have three this year?', and there are people who say they want to have this, they want this to happen. So it's symptomatic, I think, of, certainly not something that's actually been driven by us, I think, it was just something that has kind of taken over.

---

[152] The transcript of the recording of the first Genethics Club (later Forum) meeting has been deposited at the Wellcome Library, London, however public access to it is closed under the Data Protection Act. Details of associated material from the Genethics Club archive that is publicly accessible are on the Wellcome Library's catalogue: Archives and Manuscripts, reference SA/GEC. Email from Ms Zoe Fullard, Wellcome Library, to Ms Emma Jones, 11 January 2016.

[153] Professor Karen Temple has been a consultant clinical geneticist since 1990, and was instrumental in the development of the Wessex Genetics Service. She is Director of the Academic Unit of Human Development and Health at the University of Southampton; see www.southampton.ac.uk/medicine/about/staff/ikt.page (accessed 28 July 2015).

Lucassen: I think one of the things that is rather unique about the Genethics Forum is that as a medic I'm very much always driven to try and find 'the answer'. What to do, what should I do here, which is echoing what a lot of you have said: 'We were pragmatic, we just had to get on with it.' But having people from other specialties and other disciplines there to just reflect on the issues is incredibly valuable, not only as a sounding board but also to be able to construct a different way of coming with a solution than we've had any training or experience of doing, I suppose. So that's where the humanities and social science involvement in thinking about the ethical issues in genetics has been so very helpful.

Parker: Yes, I think this is quite interesting. Marcus, earlier in the day you talked about the pre-meeting, and one of the things that it's tempting to think is that the Genethics Club is something special and unique and there may be elements of that to it, but my guess is that actually there's a whole lot of discussion of this type going on in various places around the country and it's just a place where that happens in a more visible way.

Pembrey: At the risk of sticking my neck out a bit …

Lucassen: That's what we want.

Pembrey: I think what one wanted to get from this discussion and the ethical discussion, the discussion of the issues anyway and so on, is I don't know quite what it is but it's a sort of moral robustness with respect to situations that just don't fit any of the basic rules that you come across. So I suppose it's the practice and experience of framing the problem and framing the issues and coming to a practical solution. One that springs to mind particularly we saw with the neurogeneticist Michael Baraitser and others in the clinic; a woman who had come with her brother who was 22, was very severely mentally retarded, and had a very unusual face, unrecognizable syndrome and so on. And there was a cousin who was said to have the same thing and the nature of the pedigree suggested that, if those two boys were the same, it was probably X-linked. We left it at that but she rang up a week later to say that unexpectedly she was pregnant and she would like to consider a question of whether she's carrying a boy or not, and so forth. So I contacted the institution where her cousin was and they said, 'Well, we would have to get consent from the guardians and so on.' It turned out that the parents had died, the mother had died, and the guardian was the mother's brother, or ex-husband or something, and he was a belligerent person who hadn't seen this person for two years and then refused for me to go and see him. I just wanted to take a photograph, or to see him at least. And so the hospital said I can't go: those are the rules. I took [the clinical geneticist] Mike Patton along as moral support and said, 'I don't care what you say, I'm going.' You know, there's no harm to this person whatsoever and it will be vitally important to help us and the family. We went into the ward to see him; they were all terribly nervous about it. I said, 'Do you want me to sign anything?' They didn't have anything for me to sign, but it was an example of where, perhaps if we hadn't had all those discussions over a long time, one might have been kowtowed and bullied into, 'Well, if those are the rules then I'm afraid we can't help.' That's one example that springs to mind where a good training in discussing ethical issues helps one deal with such obstacles.

Turnpenny: I think the importance of our discussions, they serve a number of functions, but, most importantly, is this sense of solidarity with one another, the sense of comfort that if we come to a decision about a very difficult issue, we've shared it with people. We have more confidence in the ultimate decision we make, even if it's exactly the same decision we would have made anyway. We have some reasonable grounds for going ahead with confidence, and all that is not only important just for the way we deal with patients themselves, but it's always been important in terms of good clinical governance, simple as that. We're all meant to practise good clinical governance, and it's one of the key areas of clinical governance for us in our specialty.

On top of all of that, an issue which we are in theory vulnerable to, though doesn't happen too often, but it is actually a medico-legal kind of backup, it has a medico-legal function if we are going to either face official complaints from patients to our employing organizations, or indeed potentially be taken to court for a course of action that we've taken or the way we've dealt with patients. Clearly, if we've had those discussions in important forums like our departmental teams, the Genethics Club, or indeed other areas, then we can hand on heart say that we know the decision we reached was one that was a consensus.

Modell: Yes. Following on from what you've been saying, and looking back at what's come out at this meeting, and with particular reference to prenatal diagnoses and selective abortion, the experience of the people who were working on this early in the 1970s, I think across the board, was that this was something that was not universally accepted or approved by their peers.[154] But

---

[154] See, for example, Harris (1974).

what's happened is, of course, is that it has become progressively more accepted by society. And it is very important to formulate this in the terms that you are talking about. It's terribly important in showing the support of society for the kind of service that we offer.

Lucassen: I wonder if we want to say a little bit more about that?

Modell: The medical ethical contingent, let's say.

Lucassen: So the relationship between medical ethics and the law; people are hinting at it.

Modell: And between the individual doctor and the individual patients, and society.

Hallowell: I was just listening to what you were saying and I was thinking that, and we've touched upon it throughout the day, I wonder about the extent to which the ethical issues are now changing as genetics and genomics is now changing. So while we still have those issues around reproduction, those ethical issues, we are actually now moving towards a different set of ethical issues within genetics clinics and within the mainstreaming of genomics. I wonder how that's changing? Mike handily said there were three issues, and I think that those latter ones that you mentioned, which were the family and children, weren't they? Giving children an open future, and the reproductive issues. Now we're getting much more of a focus on the family through things like: should we disclose information that we find in genomic tests? What impact will that have on the individuals themselves and on their family members when they are not expecting those kinds of things? I wonder whether, as you rightly say, on some of the issues around reproduction maybe we've kind of come to terms with them as a society and now we're actually focusing on another set of issues, which are the issues around disclosure and incidental findings, and what we do with lots of different kinds of information that may emerge in the future.

Modell: Yes, we've come to a bit of a conclusion about the prenatal diagnosis issue but now a process exists in society for examining the dilemmas and then reaching some kind of social conclusion and I think it's a very good thing, long may it continue.

Harper: I just wanted to say a word about the last, or the bottom point, on the programme.[155] It was about the international aspects, and it is fair to say that medical genetics has always been very international, partly because it's small,

---

[155] See Table 1, page 3.

and partly because it always has been internationally focused. Bodies like the HGC and the various reports and committees, etc., they haven't just had an effect in this country, they have certainly had a lot of effect right across the whole of continental Europe and in countries like Australia and even on the other side of the Atlantic. Correspondingly, we've been influenced by some of the things from abroad. And so I think this has been a valuable effect. I'd just like to mention France where, in the course of doing my interviews, one of the people I interviewed was Jean-François Mattei, who was a clinical geneticist, head of the department in Marseille, and he was for a number of years Minister of Health in France.[156] In the early 1990s he was responsible for a bioethics law being introduced.[157] He also set up the French national ethics commission in 1983,[158] and he was able also to get the sort of issues we've been talking about today across, in terms of getting them encoded, if not in law at least in legal principles.[159] This has had a very valuable role. I don't think I know of anybody in this country who has had such political involvement. Most of us aren't particularly political in our inclination but that's very far from the case in France and, indeed, Arnold Munnich was adviser to President Sarkozy.[160] So the French have done a very good job even though it's perhaps been in a rather Napoleonic, legalistic way [laughter]. And then in a way the other extreme influence, I think it's been very important, has been the effects in Eastern Europe where, not surprisingly, given that for many years they were directed to do this and that in every aspect of their lives, and genetic counselling, such as it was, was exceptionally directive.[161] I have vivid memories of one Eastern

---

[156] A transcript of the interview with Jean-François Mattei (d. 2014) is freely available to download from the Genetics and Medicine Historical Network's website at https://genmedhist.eshg.org/fileadmin/content/website-layout/interviewees-attachments/jean-francois-mattei-interview.pdf (accessed 11 January 2016).

[157] Loi no 94-653 du 29 Juillet 1994 relative au respect du corps humain (1); available to download at www.legifrance.gouv.fr/affichTexte.do?cidTexte=JORFTEXT000000549619&categorieLien=id (accessed 28 September 2015).

[158] Comité Consultatif National d'Ethique.

[159] For a discussion of the recent history of bioethics in France, see Berthiau (2013).

[160] Professor Peter Harper also interviewed Arnold Munnich; a transcript is available on the Genetics and Medicine Historical Network with biographical details; https://genmedhist.eshg.org/fileadmin/content/website-layout/interviewees-attachments/arnold-munnich-interview.pdf (accessed 11 January 2016).

[161] See Cohen et al. (1997) for a study comparing genetic counselling practices in East and West Germany, both of which practised directive counselling methods, although they were found to be more prevalent in East Germany.

European visitor describing how she would first get the husband in the room and say, 'Well, if I were you I wouldn't marry her.' And then she'd get the wife in separately and say the same, well the other way around, anyway. I remember looking rather horrified, and she said, 'Well, that's the way things are done', but things have changed and I think again people in Eastern European countries have looked over here.

I've even been rather immodestly pleased by the fact that I think my *Practical Genetic Counselling* had a bit of an effect.[162] They printed 100,000 copies in Russia but unfortunately they hadn't signed the copyright law so I didn't get anything out of it.[163] [Laughter] But perhaps 30 years later, I had a Russian geneticist at a conference who came clutching the first edition, saying, 'I use it every day.'[164] [Laughter] And I remember looking rather alarmed and saying, 'Well, really you should get it up to date!' [Laughter] … but he seemed quite happy with it. I gave him a copy of the current edition.

So I do think there is a big international dimension to what we do both in practice and in terms of our wider discussions, and I think it's worth also looking at that perspective of it.

Hodgson: That rang a bell because about two years ago I went to India to a cancer genetics conference and people kept rushing up to me and saying, 'Can we have your rules, your protocols, your ethical advice?' So that was rather good.

Modell: Yes, the international dimension isn't limited to Europe at all. I was going to mention the example of Iran, which has a national policy about genetic counselling based on the ethical principles that were developed here, very explicitly, and using the methods which have been used in face-to-face counselling.[165]

Lucassen: It's six o'clock so I'm just going to call on people for rounding off comments now.

Pembrey: Well, in terms of the international thing, I just wanted to say that, probably, because I was Consultant Adviser to the Chief Medical Officer at the time, I was one of two people invited to the expert group of 'Advisers on

---

[162] Harper (1981), first edition; latest edition revised by Professor Angus Clarke, see Clarke (2015).

[163] Harper (1984).

[164] Professor Peter Harper wrote, 'He had travelled for three days by train from the Russian Far East to reach the meeting!' Note on draft transcript, 1 December 2014.

[165] Iran's clinical genetics services are discussed in, for example, Samavat and Modell (2004).

the Ethical Implications of Biotechnology' for the European Commission in Brussels on prenatal diagnosis, and this was a two-year cycle and then produced a document and so on.[166] And so, again, what we were developing in Britain had been fed into the European scene.

Dennis: Well, could I just volunteer that my understanding of our discussion today is, between about 1980 and 2000 there was an influx of non-medical people into genetic counselling, which coincided with, and was probably the cause of, increasing explicitness about the ethical underpinnings of our practice. I think it's been made clear that clinical genetics led the way within medicine in doing that. I hope that, and I believe that, in the same way as the development of genetic expertise could be diffused out into the rest of medicine, and indeed needed to be done so, and I think has been done; it sounds as if these ethical underpinnings are also being diffused out into the rest of medicine. I feel rather proud to be part of this specialty that seems, on the whole, to have done things rather well over the period of my career.

Turnpenny: Just very quickly on the international side. Something that does interest me, and concerns me a little bit in coming back to the Beauchamp and Childress principles, is that the justice principle always seemed to me one that is under great stress the whole time because it's all about equity of access to healthcare, and we know that is hugely problematic across the world, but it's certainly a problem for genetics, the availability of genetic advice and testing expertise, certainly laboratory testing. It's something that I think does, and should, exercise us in the ethical realm. I think it's going to become increasingly difficult for us in the immediate future, and short and medium term, given the constraints that we are going through in the health service at the moment, where our availability to be involved in other ways outside our core work is coming under increasing pressure.

Parker: I'm strongly of the view, and have written on this, that there is a sort of moral craft element to clinical genetics that is very important, and internal.[167] I do also think it's very important to emphasize, and we haven't said much about the important role played by social science, and we've got Nina and Martin here

---

[166] 'The group [of Advisers of the Ethical Implications of Biotechnology of the European Commission], which began work in 1991, is charged with finding ways to reconcile technological progress and ethical imperatives. "Its opinions help guide the European Community in its legislative and other activities"'; quoted from Group of Advisers on the Ethical Implications of Biotechnology (1996).

[167] For further discussion of 'moral craft', see Parker (2012), in particular pages 125–7, 129.

and others as well. The kind of findings, the evidence that they've found, and the analysis they've made of what's happening in the clinic and ways in which families and the public think about these things and the whole experience of being involved in genetics, along with ethics, which has tended in this area to be a sort of empirical ethics. Bobbie and I both work in that sort of area. I think that's helped. It's partly internal and it's partly about these other disciplines being drawn into a space as well.

Richards: A final reflection on that point. I've always assumed or thought what we do for a living in this field was social science, but I think at this point I realise that we've been doing ethics. I'm reflecting back to the issue you raised about what are ethics? What are principles? What are we talking about? We are talking about, I guess, a set of social relationships effectively, and it seems to me the social scientists were part of that, people like Mike, Bobbie, as well as the clinicians themselves. It's been an interesting conversation over the last 30 to 40 years.

Lucassen: Yes, and what struck me in the last few hours is our different understanding of, when we use the word 'ethics', quite what we mean by it, but I think it's been a very rich conversation and I've certainly learnt a lot. I don't know if anyone wants any last word but otherwise I suggest closing the meeting. Mike?

Parker: I just think what you said is really important because I feel that the role that we've all played, and I don't just mean the ethicists or social scientists, all of us, is to try to keep this conversation open and ongoing and alive. And that's actually why important progress has been made in this area, but it's a very creative area as well as being good for patients, hopefully, and their families.

Lucassen: That sounds like a very positive note … oh, Nick is going to follow that up.

Dennis: When I said I was proud, I didn't mean I was proud just of the medical bits, and I think one of the things I'm most proud of is the open mindedness of the medical people in clinical genetics inviting other disciplines in. I think that's been very productive.

Tansey: If I could just say thank you all very much for coming and contributing this afternoon. We will be back to all of you to ask for further elucidation and details to make this a very useful contribution to the history of the field. As I said at the beginning, nothing will be published without your permission; you wouldn't expect otherwise, would you? We are going to be ethical, [laughter] we will ask you for your copyright and if you wish to withhold any information you have every right to do that – but I hope you won't. It's been a fascinating afternoon and thank you all very much for bringing your very different contributions; I'm particularly delighted to thank you, Anneke, for your excellent chairing.

Lucassen: Thank you.

Biographical notes*

Professor Richard Ashcroft

MA(Cantab) PhD(Cantab) FHEA FRSB (b. 1969) studied mathematics before graduating in history and philosophy of science at Cambridge in 1990. He completed a PhD on ethics in science, again at Cambridge, under Professor John Forrester and Professor Nick Jardine in 1995. From 1995 to 1996 he was research fellow in the Department of Philosophy at Liverpool University, working on a study of ethics in clinical trials, funded by the NHS Health Technology Assessment programme and led by Professor Jane Hutton. He subsequently became Lecturer at the Centre for Ethics in Medicine under Professor Alastair Campbell (1997–2000) and Lecturer (and subsequently Senior Lecturer and Reader) in Biomedical Ethics at Imperial College London in the Department of Primary Care and General Practice (2000–2006). In 2006 he became Professor of Biomedical Ethics in the School of Medicine and Dentistry at Queen Mary University of London, moving to the School of Law in 2007. He has been a member of the Gene Therapy Advisory Committee, the ethics committee of the Royal College of Obstetricians and Gynaecologists, and the Ethics of Research and Patient Involvement Committee at the Medical Research Council. He is currently a member of the Tobacco Advisory Group at the Royal College of Physicians of London, the ethics advisory board of Genomics England Ltd., and the working party on genome editing at the Nuffield Council on Bioethics. He was for several years Deputy Editor of the *Journal of Medical Ethics*. He works mainly on ethics in biomedical research and public health ethics.

Dr Mark Bale

PhD (b. 1962) studied applied biology at Cardiff, and was a researcher in microbial genetics at Bristol. He joined the UK Government's Department of Health (DoH) in 1999 where he led on genetics and genomics from 2004 and was Secretary to the Human Genetics Commission. He is Deputy Head of Health Science and Bioethics in the DoH, and leads on a number of key emerging healthcare science areas and their ethical, legal, and

---

* Contributors are asked to supply details; other entries are compiled from conventional biographical sources.

policy implications. He is also Head of Profession for Scientists and Engineers, and a Deputy to the Chief Scientific Adviser, Professor Dame Sally Davies. He is currently working on the delivery of the Prime Minister's 100,000 Genomes Project, which includes the establishment of Genomics England to realise the project in close collaboration with NHS England and Health Education England. He also represents the UK on bioethics and biotechnology at the Council of Europe and the Organisation for Economic Cooperation and Development. He is Chair of the Council of Europe's Committee on Bioethics.

Professor Angus Clarke
DM FRCP FRCPCH (b. 1954) is Professor and Honorary Consultant in Clinical Genetics (Cardiff University and the Cardiff and Vale University Health Board). He graduated in genetics from Cambridge in 1976 and in clinical medicine from Oxford in 1979, and his pre-registration posts were at the John Radcliffe Hospital. He then trained in general medicine, paediatrics, and neonatal medicine before working on a research project on X-linked hypohidrotic ectodermal dysplasia with Professor Peter Harper. He moved to Newcastle and worked in both clinical genetics and paediatric neurology, establishing a genetic register for Duchenne muscular dystrophy and developing an interest in Rett syndrome. He returned to Cardiff in 1989 as Senior Lecturer in clinical genetics. Since then he has developed interests in newborn screening, genetic counselling, and in the social and ethical aspects of human genetics. He established, and directs, the Cardiff University MSc course in genetic counselling.

Dr Nick Dennis
MB BChir (Cantab) FRCP (b. 1944) trained in clinical genetics with Professor Cedric Carter at the Institute of Child Health, London (1972–1976), then worked with Professor Robin Bannerman in Buffalo, New York State (1976–1978). He was appointed Senior Lecturer in Clinical Genetics at the University of Southampton and Honorary Consultant to the Southampton University Hospitals Trust in 1978. From then until his retirement in 2007, he helped to plan and provide a clinical genetics service to the Wessex Health Region as well as participating in research and teaching in the University of Southampton.

Professor Bobbie Farsides

PhD studied at the London School of Economics, where she completed her doctoral research in 1992. She was a lecturer in the Department of Philosophy at Keele University (1986–1996), and Director of its Centre for Contemporary Ethical Studies from 1992 to 1996. She moved to the Centre of Medical Law and Ethics at King's College London as Lecturer/Senior in Medical Ethics (1996–1999; 1999–2006). She was appointed Professor of Clinical and Biomedical Ethics at the Brighton and Sussex Medical School in 2006, in which post she remains. She serves on many medical ethics-related committees, including: the British Medical Association Ethics Committee (2006–); Institute of Medical Ethics (board member/trustee 2008–); Brighton and Sussex University Hospital Trust (BSUHT) Clinical Ethics Committee (2001–); BSUHT Organ Donation Committee (2008–); UK Donation Ethics Committee (2010–); Medical Research Council Brain Bank Network Steering Committee (2010–); Emerging Science and Bioethics Advisory Committee (2012–); and Human Fertilisation and Embryology Authority's National Donation Strategy Group (2012–). Since 2013 she has been Chair of the Nuffield Council for Bioethics Working Party 'Children and Medical Research: Finding a way forward'. She is also co-editor of the Royal Society of Medicine journal *Clinical Ethics.*

Dr John Fraser Roberts

CBE DSc FRCP FRCPsych FRS (1899–1987) began research in human genetics in Edinburgh during the 1930s. After the Second World War, he was Honorary Consultant in Medical Genetics at the Royal Eastern Counties Hospital, Colchester (1946–1957). He started a genetics clinic in 1946 at the Hospital for Sick Children, Great Ormond Street, London, the first in Europe, and in 1957 the Medical Research Council created a Clinical Genetics Unit at the Institute of Child Health, Great Ormond Street, with Fraser Roberts as Director, where he remained until his retirement in 1964. He also started a genetics clinic at the Children's Hospital, Bristol, in 1957; see Pembrey (1987), Polani (1987), and Hill (1988).

Dr Alan Fryer

MD MRCP FRCP FRCPCH (b. 1954) trained at St Thomas' Hospital Medical School, University of London, and was awarded a Membership of the Royal College of Physicians (Paediatrics) in 1974. His training continued at St Thomas' Hospital, where he was house physician, then house surgeon at Burton General Hospital, Stoke-on-Trent (1979), followed by casualty officer in Sheffield Children's Hospital (1981), then Senior House Officer posts: neonatal paediatrics at Bristol Maternity Hospital (1981–1982); general paediatrics at the Royal Victoria Infirmary, Newcastle-upon-Tyne (1982); community paediatrics, Gateshead (1982–1983); and paediatric cardiology, Freeman Hospital, Newcastle-upon-Tyne (1983). He was Paediatric Registrar at Southmead Hospital, Bristol (1983–1984), and then at the Royal United Hospital, Bath (1984–1985), where he was also a Research Fellow in the Bath Unit for Research into Paediatrics (1985–1987). He became Lecturer in Medical Genetics at the University of Wales College of Medicine, Cardiff, in 1987, until his appointment as Consultant Clinical Geneticist at the Royal Liverpool University Hospital; Royal Liverpool Children's Hospital; Liverpool Women's Hospital; and the Countess of Chester Hospital, in 1990, in which post he remains. He is also Honorary Lecturer at the University of Liverpool. Since 2004 he has been an associate editor for *Archives of Diseases in Childhood,* and, since 2009, he has been a peer reviewer for pre-implantation genetic diagnosis applications on behalf of the Human Fertilisation and Embryology Authority.

Dr Nina Hallowell

BSc (Hons) DPhil MA. (b. 1957) is an applied medical sociologist who also has an MA in medical ethics and law. She has worked at Newcastle University, the Universities of Cambridge and Edinburgh, and the Institute of Cancer Research in London. She was a Reader in Social Science and Public Health at the University of Edinburgh and now works as an independent research consultant in the UK and Australia. In the mid-1990s she moved to Cambridge to work with Martin Richards on an MRC-funded project that looked at families' experiences of genetic testing for late-onset disorders, primarily breast and ovarian cancers. This project triggered a long-term interest in the impact of genetic testing on families, and she has researched this topic ever since. She has published over 60 articles in a range of medical, social science, and ethics journals. She has sat on a number of research ethics committees and is a member of the Ethics Group of the UK's National DNA Database at the Home Office.

Professor Peter Harper

Kt FRCP (b. 1939) graduated from Oxford University in 1961, qualifying in medicine in 1964. After a series of clinical posts, he trained in medical genetics at the Liverpool Institute for Medical Genetics under Cyril Clarke and at Johns Hopkins University, Baltimore, under Victor McKusick. He was Professor of Medical Genetics at the University of Wales' College of Medicine, Cardiff, from 1971 until his retirement in 2004, when he was appointed University Research Professor in Genetics, Cardiff University (Emeritus since 2008). He served on the UK's Human Genetics Commission from 2000 to 2004 and from 2004 to 2010 with the Nuffield Council on Bioethics. He has been closely involved with the identification of the genes underlying Huntington's disease and muscular dystrophies, and with their application to predictive genetic testing. He has also been responsible for the development of a general medical genetics service for Wales. His books include *Practical Genetic Counselling* (Harper, 1981); *Landmarks in Medical Genetics* (Harper (ed.), 2004); *First Years of Human Chromosomes* (Harper, 2006), and *A Short History of Medical Genetics* (Harper, 2008). For the past decade he has led an initiative, supported by the Wellcome Trust, to preserve and document the history of human and medical genetics (https://genmedhist.eshg.org/39.0.html). He is a consultant to the 'Makers of Modern Biomedicine Project' for the History of Modern Biomedicine Research Group, Queen Mary University of London.

Professor Shirley Hodgson

DM D(Obst) RCOG DCH FRCP (b. 1945), daughter of Lionel Penrose, avoided working in genetics for many years because of her father's fame, but after training in medicine and working in general practice while her children were young, she did a locum in clinical genetics at Guy's Hospital, and found the discipline irresistible. She went on to work in clinical genetics for many years. From 1983 to 1988 she was Senior Registrar in Clinical Genetics for the South Thames (East) Regional Genetics Centre and Honorary Senior Registrar at Hammersmith Hospital, London, where she did her thesis in the study of Duchenne muscular dystrophy. She then moved to Addenbrooke's Hospital, Cambridge (UK), as Consultant Clinical Geneticist (1988–1990), where she developed an interest in cancer genetics. She continued to promote this interest when she moved back to Guy's and St

Thomas' and also at St Mark's Hospital in London in the 1990s, and ran the regional cancer genetics service for the South-East Thames Region. She has published widely on the subject of cancer genetics, and co-authored several books, including *Inherited Susceptibility to Cancer* (Foulkes and Hodgson (eds) (1998)) and *A Practical Guide to Human Cancer Genetics* (Hodgson and Maher (1993)), now into its fourth edition with W Foulkes and C Eng as co-authors (Springer). She is partially retired but did have an active research programme investigating inherited aspects of cancer predisposition, especially in breast and colorectal cancers. She took up a new post as Professor of Cancer Genetics at St George's, University of London, in 2003, now Emeritus. She continues to do clinics in cancer genetics, in Leicester, and to teach in cancer genetics and global health. She has international interests, with teaching involvements in India and China, and is currently helping to develop the new medical school at the University of Namibia (UNAM).

Professor Anneke Lucassen
MD PhD (b. 1962) studied medicine at the University of Newcastle-upon-Tyne and did her clinical training in Oxford, followed by a PhD in the molecular genetics of multifactorial disease at the Institute of Molecular Medicine (Oxford). She specialized in clinical genetics, with a particular interest in cancer and cardiac genetics. She was appointed as Consultant/Senior Lecturer in Oxford in 1997, then moved to Southampton in 2000 and became Professor of Clinical Genetics in 2007. Over the past 20 years she has applied her clinical and laboratory skills to focus on the ethical issues raised as new genetic technologies are integrated into the NHS, and she now leads an interdisciplinary programme of research into the social, ethical, and legal aspects of developments in genetics through the Clinical Ethics and Law unit at Southampton. In 2001, supported by Wellcome Trust funding, she co-founded the UK Genethics Forum (previously Club), which has to date held 40 national one-day meetings to discuss ethical issues as they arise in clinical practice. She sat on the Nuffield Council of Bioethics from 2009 to 2015, and on the Human Genetics Commission until 2012. She is a member of the Ethics Advisory Group for Genomics England.

Professor Victor McKusick
MD (1921–2008) qualified in medicine at Johns Hopkins University and completed his internship and residency in internal medicine there. He was Executive

Chief of the Cardiovascular Unit at Baltimore Marine Hospital (1948–1950), while progressing through the ranks in the Johns Hopkins Department of Medicine. He also held joint professorships in epidemiology in the Johns Hopkins University School of Public Health and in biology. He founded the Division of Medical Genetics in 1957, which he headed until 1973, when he became the William Osler Professor and Chairman of the Department of Medicine, and Physician-in-Chief of Johns Hopkins Hospital. He held these posts until 1985, when he was named University Professor of Medical Genetics.

Professor Bernadette Modell

PhD FRCP FRCOG (b. 1935) undertook her first degree in zoology, predominantly in genetics and embryology, in Oxford in 1955; followed by doctoral research in developmental biology, Cambridge, 1959. She qualified in medicine at Cambridge and University College Hospital in 1964, aiming to investigate the application of genetic knowledge in medical practice. She subsequently worked at University College London and University College London Hospitals until her retirement in 2000, where her work focused on thalassaemia as an example of a common genetic

disorder. She was involved in developing effective treatment for thalassaemia major and prevention of the disease through community information; population screening and genetic counselling; and methods for prenatal diagnosis – including the first trimester diagnosis by chorionic villus sampling and DNA analysis. In collaboration with the WHO she has helped to extend programmes for the treatment and prevention of haemoglobin disorders to many parts of the world. She is currently Emeritus Professor of Community Genetics at the UCL (University College London) Centre for Health Informatics and Multiprofessional Education, where she works on the global epidemiology of congenital disorders, and developing informatics approaches to the provision of genetic information for communities. She is also Director of the WHO Collaborating Centre for Community Control of Hereditary Disorders.

Mrs Elizabeth Mumford

LLM (b. 1958) was educated at Stanford University, the University of Toronto, and Queens' College Cambridge. She was a lecturer in law at King's College London and at Bristol University. She took an early 'retirement' in 2000 to embark on a late career as a mother but still lectures on a part-time basis in medical law at Bristol.

Professor Michael Parker
BEd MA PhD (b. 1958) is Professor of Bioethics and Director of the Ethox Centre at the University of Oxford. Before becoming an academic, he worked for more than a decade with homeless teenagers in London, mostly for a charity called Centrepoint Soho. He did his undergraduate degree in education at Bristol Polytechnic (University of West of England) and a PhD in philosophy at Hull University. His first academic positions were at the University of Central Lancashire, the Open University, and Imperial College London. He moved to Oxford from Imperial College London in 1999. His overarching research interest is in the practical ethical aspects of the day-to-day work of health professionals and medical researchers, and the development of 'moral craftsmanship' in such contexts. He has a long-standing interest in the ethics of clinical genetics and, in 2001, together with Anneke Lucassen, Angus Clarke, and Tara Clancy, he established the Genethics Club – a national ethics forum for genetics professionals to identify and address ethical issues in their work. By 2015, the Genethics Club had met 40 times. This work is published as *Ethical Problems and Genethics Practice* (Parker (2012)). Michael's other main research interest is in the ethical issues arising in collaborative global health research; again focusing on day-to-day practical ethics. Together with partners in Kenya, Thailand, Vietnam, Malawi, and South Africa, he co-ordinated the Global Health Bioethics Network, which is funded by a Wellcome Trust Strategic Award.

Professor Marcus Pembrey
MD FRCP FRCOG FRCPCH FMedSci (b. 1943) is Emeritus Professor of Paediatric Genetics at the Institute of Child Health, University College London and visiting Professor of Paediatric Genetics at the University of Bristol. He graduated from Guy's Hospital in 1966 with an interest in paediatrics and medical genetics, then studied benign sickle cell in Eastern Saudi Arabia while training in clinical genetics with Paul Polani at Guy's. In 1979 he was appointed head of the new Mothercare Unit of paediatric genetics at the Institute of Child Health and honorary consultant in clinical genetics at Great Ormond Street Hospital for Children. Here he helped develop clinical DNA analysis services, contributing to the Department of Health's Special Medical

Development on this. His research focused on irregular inheritance, initially fragile X syndrome and then Angelman syndrome and genomic imprinting. This led to his current interest in transgenerational responses to early life exposures. He helped Jean Golding launch the Avon Longitudinal Study of Parents and Children (ALSPAC) in Bristol, being Director of Genetics within ALSPAC from 1989 to 2005. He was Adviser in Genetics to the Chief Medical Officer UK (1989–1998) and President of the European Society of Human Genetics (1994–1995).

Professor Martin Richards

PhD ScD (b. 1940) studied zoology at the University of Cambridge. He held a post-doctoral fellowship from the Science Research Council (1965–1967), and concurrently was research fellow at Trinity College, Cambridge (1965–1969). In 1966 he was a visiting fellow at the Department of Biology, Princeton University, and then a visitor in the Center for Cognitive Studies at Harvard University (1967–1968). He founded the Family Research Group in Cambridge, in 1967, and became its Director until 2005, when it was formally incorporated by the University. He held a series of academic posts at Cambridge University: Lecturer in Social Psychology (1970–1989);

Reader in Human Development (1989–1997); Professor of Family Research (1997–2005); Head of Department of Social and Development Psychology (2005). Since 2005 he has been Emeritus Professor of Family Research at Cambridge. He was a member of the Human Genetics Commission (1999–2005), and the Ethics and Law Advisory Committee for the Human Fertilisation and Embryology Authority (2003–2009). Currently, he is a member of the Cambridge University Hospital NHS Foundation Trust's Human Tissue Management Committee (2008–) and its Clinical Forum (2014–).

Professor Tilli Tansey

OBE PhD PhD DSc HonMD HonFRCP FMedSci (b. 1953) graduated in zoology from the University of Sheffield in 1974, and obtained her PhD in *Octopus* neurochemistry in 1978. She worked as a neuroscientist in the Stazione Zoologica Naples, the Marine Laboratory in Plymouth, the MRC Brain Metabolism Unit, Edinburgh, and was a Multiple Sclerosis Society Research Fellow at St Thomas' Hospital, London (1983–1986). After a short sabbatical break at the Wellcome Institute for the History of Medicine (WIHM), she took a second PhD in medical history on the career of Sir Henry Dale, and became a member of the academic staff of the WIHM, later the Wellcome Trust Centre for the History of Medicine at UCL. She became Professor of the History of Modern Medical Sciences at UCL in 2007 and moved to Queen Mary University of London (QMUL), with the same title, in 2010. With the late Sir Christopher Booth she created the History of Twentieth Century Medicine Group in the early 1990s, now the History of Modern Biomedicine Research Group at QMUL.

Professor Peter Douglas Turnpenny

FRCP FRCPCH FRCPath FHEA (b. 1953) is Consultant Clinical Geneticist at the Royal Devon and Exeter Healthcare NHS Trust and Honorary Associate Professor, University of Exeter Medical School. He graduated in medicine at Edinburgh University in 1977 and initially pursued a career in paediatric medicine. Between 1983 and 1990 he worked as a paediatrician (and anaesthetist) at The Nazareth Hospital, Israel (owned and governed by the Edinburgh Medical Missionary Society, now The Nazareth Trust). During the latter half of this period his interest in clinical genetics developed as he dealt with many paediatric medical conditions from a community with high levels of consanguinity. He returned to the UK in 1990 to a training position in clinical genetics in Aberdeen prior to moving to his consultant post in Exeter in 1993. As well as his interest in ethics related to medical genetics, he has clinical research interests in the genetics of spinal malformations, fetal anticonvulsant syndromes, and hypermobility. Since 2004 he has been the lead author of the award-winning *Emery's Elements of Medical Genetics*, whose 14th edition was published in 2011 (Turnpenny (2012)). From 2011 to 2013 he was President of the Clinical Genetics Society (UK). He has also been a Trustee of The Nazareth Trust since 2004 and helps Palestinians in their aspirations to develop genetic services. In April 2015 he was awarded an Honorary Clinical Professorship by the University of Exeter Medical School.

References[*]

Advisory Committee on Genetic Testing. (1997) *Code of Practice and Guidance on Human Genetic Testing Services Supplied Direct to the Public.* London: Health Departments of the United Kingdom.

Ahmed A, Green J, Hewison J. (2002) What are Pakistani women's experiences of antenatal carrier screening for beta-thalassaemia in the UK? Why it is so difficult to answer this question? *Public Health* **116:** 297–9.

Albert B. (2007) Disability rights, genetics and public health. In Douglas J, Earle S, Handsley S *et al.* (eds) *A Reader in Promoting Public Health: Challenge and controversy.* London; Milton Keynes: Sage/Open University: 75–81.

Alderson P, Williams C, Farsides B. (2001) *Cross currents in Genetics and Ethics around the Millennium 1999–2001.* London: Institute of Education.

Almond B, Hill D. (eds) (1991) *Applied Philosophy: Morals and metaphysics in contemporary debate.* London: Routledge.

Anon. (1999) Sweden to compensate sterilised women. 4 March, BBC News: http://news.bbc.co.uk/1/hi/health/background_briefings/international/290661.stm (accessed 8 October 2015).

Ashcroft R, Lucassen A, Parker P *et al.* (eds) (2005) *Case Analysis in Clinical Ethics.* Cambridge: Cambridge University Press.

Balint M. (1957) *The Doctor, his Patient and the Illness.* London: Pitman Medical.

Beauchamp T L, Childress J F. (1979) *Principles of Biomedical Ethics.* New York, NY; Oxford: Oxford University Press.

Bechtel K, Geschwind M D. (2013) Ethics in prion disease. *Progress in Neurobiology* **110:** doi: 10.1016/j.pneurobio.2013.07.001.

Bell A. (2015) Science for the People. *Mosaic* **January:** available at http://mosaicscience.com/story/science-people (accessed 5 January 2016).

---

[*] Please note that references with four or more authors are cited using the first three names followed by 'et al.'. References with 'et al.' are organized in chronological order, not by second author, so as to be easily identifiable from the footnotes.

Bell J. (1934–1947) *Treasury of Human Inheritance.* Four volumes. (ed. Fisher R A, clinical notes by Purdon Martin J.) London: Cambridge University Press.

Berthiau D. (2013) Law, bioethics and practice in France: forging a new legislative pact. *Medicine, Health Care and Philosophy* **16:** 105–13.

Borry P, Evers-Kiebooms G, Cornel M C *et al.* (2009) Genetic testing in asymptomatic minors: Background considerations towards ESHG recommendations. *European Journal of Human Genetics* **17:** 711–19.

Bowles Biesecker B, Marteau T M. (1999) The future of genetic counselling: an international perspective. *Nature Genetics* **22:** 133–7.

Campbell A V. (1972) *Moral Dilemmas in Medicine: A coursebook in ethics for doctors and nurses.* Edinburgh: Churchill Livingstone.

Carter C O, Evans K A, Fraser Roberts J A *et al.* (1971) Genetic clinic: A follow-up. *Lancet* **297:** 281–5.

Christie D A, Tansey E M. (2003) *Genetic Testing.* Wellcome Witnesses Twentieth Century Medicine, vol. 17. London: Wellcome Trust Centre for the History of Medicine at UCL, freely available to download at www.histmodbiomed.org/witsem/vol17 (accessed 7 July 2015).

Ciba Foundation. (1985) *Abortion: Medical progress and social implications.* Symposium on Abortion. London: Pitman.

Clarke A [J]. (1994) The genetic testing of children. Report of a Working Party of the Clinical Genetics Society (UK). *Journal of Medical Genetics* **31:** 785–97.

Clarke A J. (1997a) The process of genetic counselling: Beyond non-directiveness. In Harper P S, Clarke A J. *Genetics, Society and Clinical Practice.* Oxford: BIOS.

Clarke A J. (1997b) Prenatal genetic screening. In Harper P S, Clarke A. (eds) *Genetics, Society and Clinical Practice.* Oxford: BIOS; 119–40.

Clarke A J. (1997c) The genetic testing of children. In Harper P S, Clarke A J. *Genetics, Society and Clinical Practice.* Oxford: BIOS; 15–30.

Clarke A [J]. (2015) *Harper's Practical Genetic Counselling.* London: Taylor & Francis.

Clarke C A. (1968) The prevention of 'Rhesus' babies. *Scientific American* **219**: 46–52.

Clarke C A. (1972) Prevention of Rh isoimmunization. *Progress in Medical Genetics* 8: 169–223.

Cohen P E, Wertz D C, Nipperts I *et al.* (1997) Genetic counseling practices in Germany: A comparison between East German and West German geneticists. *Journal of Genetic Counseling* **6:** 61–80.

Committee of Inquiry into Human Fertilisation and Embryology, Warnock M. (1984) *Report of the Committee of Inquiry into Human Fertilisation and Embryology.* Cmnd 9314. London: HMSO.

Cuckle H S. (1994) Screening for neural tube defects. *Ciba Foundation Symposium* **181:** 253–66.

Department of Health. (1993) *Population Needs and Genetic Services: An outline guide.* London: HMSO.

Department of Health. (2003) *Our Inheritance, Our Future: Realising the potential of genetics in the NHS.* Cmnd 5791. London: The Stationery Office.

Department of Trade and Industry. (1996a) *Human Genetics: The Science and its Consequences. Government response to the Third Report of the House of Commons Select Committee on Science and Technology 1994–95 session.* Cmnd 3061. London: HMSO.

Department of Trade and Industry. (1996b) *Government Response to the House of Commons Science and Technology Committees Report on Human Genetics.* Cmnd 3306. London: HMSO.

van Dijk A, Oten W, Zoeteweij M W *et al.* (2003) Genetic counselling and the intention to undergo prophylactic mastectomy: effects of a breast cancer risk assessment. *British Journal of Cancer* **88:** 1675–81.

Emanuel E J, Grady C J, Crouch R A *et al.* (eds) (2008) *The Oxford Textbook of Clinical Research Ethics.* Oxford: Oxford University Press.

Emery J, Hayflick S. (2001) The challenge of integrating genetic medicine into primary care. *British Medical Journal* **322:** 1027–30.

European Community Huntington's Disease Collaborative Study Group. (1993) Ethical and social issues in presymptomatic testing for Huntington's disease: a European Community collaborative study. *Journal of Medical Genetics* **30**: 1028–35.

European Society of Human Genetics. (2009) Genetic testing in asymptomatic minors: Recommendations of the European Society of Human Genetics. *European Journal of Human Genetics* **17**: 720–1.

Evans D G, Barwell J, Eccles D M *et al.* (2014) The Angelina Jolie effect: how high celebrity profile can have a major impact on provision of cancer related services. *Breast Cancer Research* **16**: 442.

Evers-Kiebooms G, Cassiman J J, van den Berghe H. (1987) Attitudes towards predictive testing in Huntington's disease: a recent survey in Belgium. *Journal of Medical Genetics* **24**: 275–9.

Farsides B. (2011) Courage, compassion and communication: young people and Huntington's disease. *Clinical Ethics* **6**: 55.

Fishel R, Lescoe M K, Rao M R S *et al.* (1993) The human mutator gene homolog *MSH2* and its association with hereditary nonpolyposis colon cancer. *Cell* **75**: 1027–38.

Fletcher J C, Berg K, Tranøy K E. (1985) Ethical aspects of medical genetics. A proposal for guidelines in genetic counseling, prenatal diagnosis, and screening. *Clinical Genetics* **27**: 199–205.

Foulkes W D, Hodgson S V. (eds) (1998) *Inherited Susceptibility to Cancer: Clinical, predictive, and ethical perspectives.* Cambridge; New York: Cambridge University Press.

Fraser F C. (1961) On being a medical geneticist. *American Journal of Human Genetics* **15**: 1–10.

Fraser Roberts J A. (1940) *An Introduction to Medical Genetics.* Oxford: Oxford University Press.

Fraser Roberts J A, Pembrey M E. (1978) *An Introduction to Medical Genetics.* Oxford: Oxford University Press.

French Anderson W. (1989) Human gene therapy: Why draw a line? *Journal of Medicine and Philosophy* **14**: 681–93.

General Medical Council, Education Committee. (1993) *Tomorrow's Doctors: Recommendations on undergraduate medical education.* London: General Medical Council.

Gillon R. (1985) *Philosophical Medical Ethics.* Chichester; New York: John Wiley.

Groden J, Thliveris A, Samowitz W *et al.* (1991) Identification and characterization of the familial adenomatous polyposis coli gene. *Cell* **66:** 589–600.

Group of Advisers on the Ethical Implications of Biotechnology. (1996) Ethical aspects of prenatal diagnosis. *Politics and the Life Sciences* **15:** 329–34.

Gusella J F, Wexler N S, Conneally P M *et al.* (1983) A polymorphic DNA marker genetically linked to Huntington's disease. *Nature* **306:** 234–8.

Hallowell N, Murton F, Statham H *et al.* (1997) Women's need for information before attending genetic counselling for familial breast or ovarian cancer: a questionnaire, interview and observational study. *British Medical Journal* **314:** 281–3.

Hallowell N, Ardern-Jones A, Eeles R *et al.* (2005) Communication about genetic testing in families of male *BRCA1/2* carrier and non-carriers: Patterns, priorities and problems. *Clinical Genetics* **67:** 492–502.

Hanson C. (2013) *Eugenics, Literature, and Culture in Post-War Britain.* London; New York, NY: Routledge.

Harper K, Winter R M, Pembrey M E *et al.* (1984) A clinically useful DNA probe closely linked to haemophilia A. *Lancet* **324:** 6–8.

Harper P S. (1981) *Practical Genetic Counselling,* 1st edn. Bristol: John Wright.

Harper P S. (1984) *Prakticheskoe Mediko-geneticheskoe Konsul'tirovanie.* Moskva: Meditsina.

Harper P S. (ed.) (2004) *Landmarks in Medical Genetics: Classic papers with commentaries.* Oxford: Oxford University Press.

Harper P S. (2006) *First Years of Human Chromosomes: The beginnings of human cytogenetics.* Bloxham: Scion.

Harper P S. (2008) *A Short History of Medical Genetics.* New York: Oxford University Press.

Harper P S. (2010) *Practical Genetic Counselling*, 7th edn. London: Hodder Arnold.

Harper P S, Clarke A. (1990) Should we test children for "adult" genetic diseases? *Lancet* **335:** 1205–6.

Harper P S, Morris M, Tyler A. (1991) Predictive tests in Huntington's disease. In Harper P S. (ed.) *Huntington's Disease.* London: Saunders; 373–413.

Harper P S, Reynolds L A, Tansey E M. (eds) (2010) *Clinical Genetics in Britain: Origins and development.* Wellcome Witnesses to Twentieth Century Medicine, vol. 39. London: Wellcome Trust Centre for the History of Medicine at UCL; freely available to download from www.histmodbiomed.org/witsem/vol39 (accessed 25 June 2015).

Harris H. (1973) Lionel Sharples Penrose. 1989–1972. *Biographical Memoirs of Fellows of the Royal Society* **19:** 521–61.

Harris H. (1974) *Prenatal Diagnosis and Selective Abortion.* London: Nuffield Provincial Hospitals Trust.

Hawkes N. (1997) Code of practice to regulate sale of DIY gene tests. *The Times,* 24 September, page 6.

Heimler A. (1997) An oral history of the National Society of Genetic Counselors. *Journal of Genetic Counseling* **6:** 315–36.

Hill A B. (1988) Obituary: Dr J A Fraser Roberts. *Journal of the Royal Statistical Society Series A (Statistics in Society)* **151:** 359–60.

Hodgson S V, Maher E R. (1993) *A Practical Guide to Human Cancer Genetics.* Cambridge: Cambridge University Press.

Hodgson S, Milner B, Brown I *et al.* (1999) Cancer genetics services in Europe. *Disease Markers* **15:** 3–13.

Hope R A, Fulford K W M, Yates A. (1996) *Oxford Practice Skills Course: Ethics, law and communication skills in healthcare education.* Oxford: Oxford University Press.

Hostetler J A. (1980) *Amish Society.* Baltimore; London: Johns Hopkins University Press.

House of Commons Select Committee on Science and Technology. (1996) *Human Genetics: The science and its consequences.* Vols 1–4. Cmnd 3061. London: HMSO.

Human Genetics Commission. (2000) *Whose Hands on Your Genes? A discussion document on the storage, protection and use of personal genetic information.* London: Human Genetics Commission.

Human Genetics Commission. (2002) *Inside Information: Balancing interests in the use of personal genetic data.* London: Department of Health.

Human Genetics Commission. (2006) *Making Babies: Reproductive decisions and genetic technologies.* London: Human Genetics Commission.

Huntington's Disease Collaborative Research Group. (1993) A novel gene containing a trinucleotide repeat that is expanded and unstable on Huntington's disease chromosomes. *Cell* **72:** 971–83.

Illes J, Sahakian B J. (2011) *Oxford Handbook of Neuroethics.* Oxford: Oxford University Press.

Joint Working Group of the Human Genetics Commission and the UK National Screening Committee. (2005) *Profiling the Newborn: A prospective technology?* London: Human Genetics Commission.

Jones E M, Tansey E M. (eds) (2013) *Clinical Cancer Genetics: Polyposis and familial colorectal cancer c.1975–c.2010.* Wellcome Witnesses to Contemporary Medicine, vol. 46. London: Queen Mary University of London; available to download at www.histmodbiomed.org/witsem/vol46 (accessed 23 June 2015).

Jones E M, Tansey E M. (eds) (2014) *Clinical Molecular Genetics in the UK c.1975–c.2000.* Welllcome Witnesses to Contemporary Medicine, vol. 48. London: Queen Mary University of London; freely available at www.histmodbiomed.org/witsem/vol48 (accessed 6 October 2015).

Jones E M, Tansey E M. (eds) (2015) *The Development of Narrative Practices in Medicine c.1960–c.2000.* Wellcome Witnesses to Contemporary Medicine, vol. 52. London: Queen Mary University of London; available to download from www.histmodbiomed.org/witsem/vol52 (accessed 12 October 2015).

Jones J S. (1993) The Galton Laboratory. In Keynes W M. (ed.) *Sir Francis Galton: Legacy of his ideas.* Basingstoke, Hampshire: Macmillan.

Kevles D J. (1985) *In the Name of Eugenics: Genetics and the uses of human heredity.* New York: Alfred A Knopf.

Leach F S, Nicolaides N C, Papadopoulos N *et al.* (1993) Mutations of a *mutS* homolog in hereditary nonpolyposis colorectal cancer. *Cell* **75:** 1215–25.

Lockwood M. (ed.) (1985) *Moral Dilemmas in Modern Medicine.* Oxford: Oxford University Press.

Lucassen A, Parker M. (2006) The UK Genethics Club: Clinical ethics support for genetic services. *Clinical Ethics* **1:** 219–23.

Manjoney D M, McKegney F P. (1978–1979) Individual and family coping with polycystic kidney disease: the harvest of denial. *International Journal of Psychiatry in Medicine* **9:** 19–31.

Manson N C, O'Neill O. (2007) *Rethinking Informed Consent in Bioethics.* Cambridge: Cambridge University Press.

Markel H. (1992) The stigma of disease: Implications of genetic screening. *American Journal of Medicine* **93:** 209–15.

Marteau T, Richards M. (1996) *The Troubled Helix: Social and psychological implications of the new human genetics.* Cambridge; New York: Cambridge University Press.

Mason J K, McCall Smith A. (1983) *Law and Medical Ethics.* London: Butterworths.

McCarthy Veach P, LeRoy B S, Bartels D M. (2003) *Facilitating the Genetic Counseling Process.* New York, NY: Springer.

McKusick V A. (1978) *Medical Genetic Studies of the Amish. Selected Papers.* Baltimore: Johns Hopkins University Press.

McKusick V A, Kelly T E, Dorst J P. (1973) Observations suggesting allelism of the achondroplasia and hypochondroplasia genes. *Journal of Medical Genetics* **10:** 11–16.

Miki Y, Swensen J, Shattuck-Eidens D *et al.* (1994) A strong candidate for the breast and ovarian cancer susceptibility gene *BRCA1. Science* **266:** 66–71.

Minna Stern A. (2012) *Telling Genes: The story of genetic of counseling in America.* Baltimore, MD: Johns Hopkins University Press.

Modell B, Berdoukas V. (1984) *The Clinical Approach to Thalassaemia.* London: Grune and Stratton.

Modell B, Ward R H T, Fairweather. (1980) Effect of introducing antenatal diagnosis on reproductive behaviour of families at risk for thalassaemia major. *British Medical Journal* **280:** 1347–50.

Nuffield Council on Bioethics. (1993) *Genetic Screening: Ethical issues.* London: Nuffield Foundation.

Nuffield Council on Bioethics. (1995) *Human Tissues: Ethical and legal issues.* London: Nuffield Council on Bioethics.

Nuffield Council on Bioethics. (2011) *Human Bodies: Donation for medicine and research.* London: Nuffield Council on Bioethics.

Ogino S, Wilson R B. (2004) Bayesian analysis and risk assessment in genetic counseling and testing. *Journal of Molecular Diagnostics* **6:** 1–9.

O'Neill O. (2002) *Autonomy and Trust in Bioethics.* Cambridge; New York: Cambridge University Press.

Overy C, Reynolds L A, Tansey E M. (eds) (2012) *History of the Avon Longitudinal Study of Parents and Children (ALSPAC), c. 1980–2000.* Wellcome Witnesses to Twentieth Century Medicine, vol. 44. London: Queen Mary University of London; freely available to download from www.histmodbiomed.org/witsem/vol44 (accessed 30 June 2015).

Pappworth M H. (1967) *Human Guinea Pigs: Experimentation on man.* London: Routledge & Kegan Paul.

Parker M. (2012) *Ethical Problems and Genetics Practice.* Cambridge: Cambridge University Press.

Pembrey M E. (1987) Obituary: Dr John Alexander Fraser Roberts. *Journal of Medical Genetics* **24**: 442–4.

Pembrey M E, Anionwu E N. (1996) Ethical aspects of genetic screening and diagnosis. In Emery A E, Rimoin D L, Pyeritz E, Connor JM (eds*). Principles and Practice of Medical Genetics,* 3rd edn. New York: Churchill Livingstone.

Penrose L S. (1946) Phenylketonuria – a problem in eugenics. *Lancet* **247:** 949–53, reprinted in *Annals of Human Genetics* **62:** 193–202.

Penrose L S. (1949) *The Biology of Mental Defect.* Preface by J B S Haldane. London: Sidgwick & Jackson.

Penrose L S. (1950) Propagation of the unfit. *Lancet* **256**: 425–7.

Penrose L S, Smith G F. (1966) *Down's Anomaly.* London: J & A Churchill.

Polani P E. (1987) John Alexander Fraser Roberts. *Biographical Memoirs of Fellows of the Royal Society* **38**: 307–22.

Reed S C. (1955) *Counseling in Medical Genetics.* Philadelphia, PA: Saunders.

Reynolds L A, Tansey E M. (eds) (2007) *Medical Ethics Education in Britain, 1963–1993.* Wellcome Witnesses to Twentieth Century Medicine, vol. 31. London: Wellcome Trust Centre for the History of Medicine at UCL; freely available to download at www.histmodbiomed.org/witsem/vol31 (accessed 4 June 2015).

Richards M. (2016) The development of governance and regulation of donor conception in the UK. In Golombok S, Scott R, Wilkinson S *et al.* (eds) *The Regulation of Reproductive Donation.* Cambridge; New York: Cambridge University Press.

Roll-Hansen N. (1989) Geneticists and the eugenics movement in Scandinavia. *British Journal for the History of Science* **22**: 335–46.

Royal College of Physicians. (1994) *Independent Ethical Review of Studies Involving Personal Medical Records: Report of a working group to the Royal College of Physicians Committee on Ethical Issues in Medicine.* London: Royal College of Physicians.

Samavat A, Modell B. (2004) Iranian national thalassaemia screening programme. *British Medical Journal* **329**: 1134–7.

Sarkar S. (1992) Science, philosophy, and politics in the work of J. B. S. Haldane. *Biology and Philosophy* **7**: 385–409.

Select Committee on Science and Technology. (2001) *Fifth Report: Genetics and insurance.* London: The Stationery Office Ltd, available online at www.publications.parliament.uk/pa/cm200001/cmselect/cmsctech/174/17404.htm (accessed 11 August 2015).

Shotter E F. (2004) Obituary: Professor the Reverend Canon G R Dunstan. *Journal of Medical Ethics* **30**: 233–4.

Skirton H, Barnes C, Kershaw A *et al.* (1998) Recommendations for education and training of genetic nurses and counsellors in the United Kingdom. *Journal of Medical Genetics* **35:** 410–12.

Streetly A, Latinovic R, Henthorn J. (2010) Positive screening and carrier results for the England-wide universal newborn sickle cell screening programme by ethnicity and area for 2005–07. *Journal of Clinical Pathology* **63:** 626–9.

Taylor A J, Lloyd J. (1995) The role of the Gene Therapy Advisory Committee in the oversight of gene therapy research in the United Kingdom. *Biologicals* **23:** 37–8.

Turnpenny P D. (2012) *Emery's Elements of Medical Genetics.* 14th edition. Philadelphia, PA: Elsevier/Churchill Livingstone.

Tyler A, Harper P S. (1983) Attitudes of subjects at risk and their relatives towards genetic counselling in Huntington's chorea. *Journal of Medical Genetics* **20:** 179–88.

Tyler A, Ball D, Craufurd D. (1992) Presymptomatic testing for Huntington's disease in the United Kingdom. *British Medical Journal* **304:** 1593–6.

Wellcome Trust Policy Unit. (*c.*2001) *Review of the Wellcome Trust Biomedical Ethics Programme.* London: Wellcome Trust.

Wertz D C, Fletcher J C. (1993) Geneticists approach ethics: an international survey. *Clinical Genetics* **43:** 104–10.

Williams C, Alderson P, Farsides B. (2002a) Dilemmas encountered by health practitioners offering nuchal translucency screening: a qualitative case study. *Prenatal Diagnosis* **22:** 216–20.

Williams C, Alderson P, Farsides B. (2002b) What constitutes balanced information in practitioners' portrayals of Down's syndrome? *Midwifery* **18:** 230–7.

Williams C, Alderson P, Farsides B. (2002c) Is nondirectiveness possible within the context of antenatal screening and testing? *Social Science and Medicine* **54:** 339–47.

Williams C, Alderson P, Farsides B. (2002d) Too many choices? Hospitals and community staff reflect on the future of prenatal screening. *Social Science and Medicine* **55:** 743–53.

Williams C, Alderson P, Farsides B. (2002e) 'Drawing the line' in prenatal screening and testing: health practitioners' discussions. *Health, Risk and Society* **4**: 61–75.

Winter R M, Harper K, Goldman E *et al.* (1985) First trimester prenatal diagnosis and detection of carriers of haemophilia A using the linked DNA probe DX13. *British Medical Journal* **291**: 765–9.

Wolstenholme G. (ed.) (1989) Cedric Oswald Carter. *Munks Roll* **8**: 78–80.

World Health Organization. (1985) *Community Approaches to the Control of Hereditary Diseases*. Report of a WHO Advisory Group. Geneva: World Health Organization, WHO/HGN/WG/85.10; available to download from www.who.int/genomics/publications/reports/en/ (accessed 12 October 2015).

Zallen D T, Christie D A, Tansey E M. (eds) (2004) *The Rhesus Factor and Disease Prevention*. Wellcome Witnesses to Twentieth Century Medicine, vol. 22. London: Wellcome Trust Centre for the History of Medicine at UCL; freely available to download from www.histmodbiomed.org/witsem/vol22 (accessed 28 July 2015).

Index: Subject

23andMe, 70
100,000 Genomes Project, 68

abortion *see* termination of pregnancy
access to genetic services, 79
    ethnic minorities, 55
activist groups, xiii, 57, 65
Addenbrookes Hospital (Cambridge),
    29, 33
Advisers on the Ethical Implications
    of Biotechnology (European
    Commission group), 79
Advisory Committee on Genetic
    Testing (ACGT), 67, 67–8,
    69–70
Agriculture and Environment
    Biotechnology Commission
    (AEBC), 67
alpha-fetoprotein, 11
Amish, 16
antenatal clinics, 28–9, 71
    see also prenatal screening and
        diagnosis
assisted reproduction, 21
autonomy, 15, 18, 20, 21
    of children, 45
Avon Longitudinal Study of Pregnancy
    and Childhood (ALSPAC), xiv,
    61–3

bad news, giving to patients/families,
    56, 57, 59, 64
Bayesian analysis, 47
bioethics *see* medical ethics
BRCA genes, 29, 33, 39
breast cancer, 29–31, 39
British Paediatric Association, 23

Canada, 7
cancer
    diagnosis and genetic testing, 30
    disclosure of test results, 39–40
    discussion groups, 41
    family cancer clinics (research), 29,
        33–4
    genes conferring susceptibility, 33
    genetic counselling, 29–31, 39, 53
    treatment/prophylaxis, 30
Cardiff University, 40, 49, 50, 51
carrier screening, 35, 46, 47, 55, 70
Case Analysis in Clinical Ethics
    (Ashcroft), xi
Centre for Family Research
    (Cambridge), 33
Cesagen, 40
children
    attitude towards affected, 9–10
    genetic testing, 44, 45–6, 72
Ciba Foundation meetings, 12–13,
    21–2
communication
    of bad news, 56, 57, 59, 64
    listening to patients, 58
    within families, 37, 38, 39–40, 72,
        74–5, 76
confidentiality, 21, 76
    in the ALSPAC study, 61–2
    in Huntington's disease testing, 37,
        38
consent
    for access to a patient, 74–5
    in Huntington's disease testing, 37,
        38
    for using clinical samples in research,
        xiv, 62–3
contraception, 18

Index: Names

Biographical notes appear in bold

VOLUMES IN THIS SERIES*

1. **Technology transfer in Britain: The case of monoclonal antibodies**
   Self and non-self: A history of autoimmunity
   Endogenous opiates
   The Committee on Safety of Drugs (1997) ISBN 1 86983 579 4

2. **Making the human body transparent: The impact of NMR and MRI**
   Research in general practice
   Drugs in psychiatric practice
   The MRC Common Cold Unit (1998) ISBN 1 86983 539 5

3. **Early heart transplant surgery in the UK (1999)** ISBN 1 84129 007 6

4. **Haemophilia: Recent history of clinical management (1999)**
   ISBN 1 84129 008 4

5. **Looking at the unborn: Historical aspects of**
   obstetric ultrasound (2000) ISBN 1 84129 011 4

6. **Post penicillin antibiotics: From acceptance to resistance? (2000)**
   ISBN 1 84129 012 2

7. **Clinical research in Britain, 1950–1980 (2000)**
   ISBN 1 84129 016 5

8. **Intestinal absorption (2000)**
   ISBN 1 84129 017 3

9. **Neonatal intensive care (2001)**
   ISBN 0 85484 076 1

10. **British contributions to medical research and education in Africa**
    after the Second World War (2001) ISBN 0 85484 077 X

* All volumes are freely available online at: www.histmodbiomed.org/article/wellcome-witnesses-volumes . They can also be purchased from www.amazon.co.uk; www.amazon.com, and from all good booksellers.

11. **Childhood asthma and beyond (2001)**
    ISBN 0 85484 078 8

12. **Maternal care (2001)**
    ISBN 0 85484 079 6

13. **Population-based research in south Wales: The MRC Pneumoconiosis Research Unit and the MRC Epidemiology Unit (2002)**
    ISBN 0 85484 081 8

14. **Peptic ulcer: Rise and fall (2002)**
    ISBN 0 85484 084 2

15. **Leukaemia (2003)**
    ISBN 0 85484 087 7

16. **The MRC Applied Psychology Unit (2003)**
    ISBN 0 85484 088 5

17. **Genetic testing (2003)**
    ISBN 0 85484 094 X

18. **Foot and mouth disease: The 1967 outbreak and its aftermath (2003)**
    ISBN 0 85484 096 6

19. **Environmental toxicology: The legacy of *Silent Spring* (2004)**
    ISBN 0 85484 091 5

20. **Cystic fibrosis (2004)**
    ISBN 0 85484 086 9

21. **Innovation in pain management (2004)**
    ISBN 978 0 85484 097 7

22. **The Rhesus factor and disease prevention (2004)**
    ISBN 978 0 85484 099 1

23. **The recent history of platelets in thrombosis and other disorders (2005)** ISBN 978 0 85484 103 5

24. **Short-course chemotherapy for tuberculosis (2005)**
ISBN 978 0 85484 104 2

25. **Prenatal corticosteroids for reducing morbidity and mortality after preterm birth (2005)** ISBN 978 0 85484 102 8

26. **Public health in the 1980s and 1990s: Decline and rise? (2006)**
ISBN 978 0 85484 106 6

27. **Cholesterol, atherosclerosis and coronary disease in the UK, 1950–2000 (2006)** ISBN 978 0 85484 107 3

28. **Development of physics applied to medicine in the UK, 1945–1990 (2006)** ISBN 978 0 85484 108 0

29. **Early development of total hip replacement (2007)**
ISBN 978 0 85484 111 0

30. **The discovery, use and impact of platinum salts as chemotherapy agents for cancer (2007)** ISBN 978 0 85484 112 7

31. **Medical ethics education in Britain, 1963–1993 (2007)**
ISBN 978 0 85484 113 4

32. **Superbugs and superdrugs: A history of MRSA (2008)**
ISBN 978 0 85484 114 1

33. **Clinical pharmacology in the UK, _c._ 1950–2000: Influences and institutions (2008)** ISBN 978 0 85484 117 2

34. **Clinical pharmacology in the UK, _c._ 1950–2000: Industry and regulation (2008)** ISBN 978 0 85484 118 9

35. **The resurgence of breastfeeding, 1975–2000 (2009)**
ISBN 978 0 85484 119 6

36. **The development of sports medicine in twentieth-century Britain (2009)** ISBN 978 0 85484 121 9

37. **History of dialysis, *c.*1950–1980 (2009)** ISBN 978 0 85484 122 6

38. **History of cervical cancer and the role of the human papillomavirus, 1960–2000 (2009)** ISBN 978 0 85484 123 3

39. **Clinical genetics in Britain: Origins and development (2010)**
    ISBN 978 0 85484 127 1

40. **The medicalization of cannabis (2010)**
    ISBN 978 0 85484 129 5

41. **History of the National Survey of Sexual Attitudes and Lifestyles (2011)** ISBN 978 0 90223 874 9

42. **History of British intensive care, *c.*1950–*c.*2000 (2011)**
    ISBN 978 0 90223 875 6

43. **WHO Framework Convention on Tobacco Control (2012)**
    ISBN 978 0 90223 877 0

44. **History of the Avon Longitudinal Study of Parents and Children (ALSPAC), *c.*1980–2000 (2012)**
    ISBN 978 0 90223 878 7

45. **Palliative medicine in the UK *c.*1970–2010 (2013)**
    ISBN 978 0 90223 882 4

46. **Clinical cancer genetics: Polyposis and familial colorectal cancer *c.*1975–*c.*2010 (2013)** ISBN 978 0 90223 885 5

47. **Drugs affecting 5-HT systems (2013)**
    ISBN 978 0 90223 887 9

48. **Clinical molecular genetics in the UK *c.*1975–*c.*2000 (2014)**
    ISBN 978 0 90223 888 6

49. **Migraine: Diagnosis, treatment and understanding *c.*1960–2010 (2014)**
    ISBN 978 0 90223 894 7

UNPUBLISHED WITNESS SEMINARS

1994   **The early history of renal transplantation**

1994   **Pneumoconiosis of coal workers**
(partially published in volume 13, *Population-based research in South Wales*)

1995   **Oral contraceptives**

2003   **Beyond the asylum: Anti-psychiatry and care in the community**

2003   **Thrombolysis**
(partially published in volume 27, *Cholesterol, atherosclerosis and coronary disease in the UK, 1950–2000*)

2007   **DNA fingerprinting**

The transcripts and records of all Witness Seminars are held in archives and manuscripts, Wellcome Library, London, at GC/253.

OTHER PUBLICATIONS

Technology transfer in Britain: The case of monoclonal antibodies
Tansey E M, Catterall P P. (1993) *Contemporary Record* **9**: 409–44.

Monoclonal antibodies: A witness seminar on contemporary medical history
Tansey E M, Catterall P P. (1994) *Medical History* **38**: 322–7.

Chronic pulmonary disease in South Wales coalmines: An eye-witness account of the MRC surveys (1937–42)
P D'Arcy Hart, edited and annotated by E M Tansey. (1998)
Social History of Medicine **11**: 459–68.

Ashes to Ashes – The history of smoking and health
Lock S P, Reynolds L A, Tansey E M. (eds) (1998) Amsterdam: Rodopi BV,
228pp. ISBN 90420 0396 0 (Hfl 125) (hardback). Reprinted 2003.

Witnessing medical history. An interview with Dr Rosemary Biggs
Professor Christine Lee and Dr Charles Rizza (interviewers). (1998)
Haemophilia **4**: 769–77.

Witnessing the Witnesses: Pitfalls and potentials of the Witness Seminar in twentieth century medicine
Tansey E M, in Doel R, Søderqvist T. (eds) (2006) *Writing Recent Science: The historiography of contemporary science, technology and medicine.* London: Routledge: 260–78.

The Witness Seminar technique in modern medical history
Tansey E M, in Cook H J, Bhattacharya S, Hardy A. (eds) (2008) *History of the Social Determinants of Health: Global Histories, Contemporary Debates.* London: Orient Longman: 279–95.

Today's medicine, tomorrow's medical history
Tansey E M, in Natvig J B, Swärd E T, Hem E. (eds) (2009) *Historier om helse* (*Histories about Health,* in Norwegian). Oslo: *Journal of the Norwegian Medical Association:* 166–73.

Key to cover photographs

Front cover, left to right
Professor Peter Harper
Professor Peter Turnpenny
Professor Anneke Lucassen
Professor Angus Clarke

Back cover, left to right
Dr Nick Dennis
Dr Nina Hallowell
Professor Martin Richards
Professor Michael Parker

Lightning Source UK Ltd.
Milton Keynes UK
UKOW07f1942260216

269201UK00001B/25/P

9 781910 195130